全国高等医药院校药学类实验教材

药物合成反应实验

（第三版）

主　编　翟　鑫
副主编　刘　洋
编　者　（以姓氏笔画为序）
王　钝　付　晔　刘　洋
沙　宇　郭　春　翟　鑫

中国健康传媒集团
中国医药科技出版社

内容提要

本教材为“全国高等医药院校药学类实验教程”之一，是在2014年出版的《药物合成反应实验（第二版）》的基础上为更好地满足实验教学需求修订而成。本教材着眼于“药物合成反应”理论课教学中重点介绍的九大反应类型，针对性地安排了23个（次）实验内容，构建了基础性实验、综合性实验、设计性实验三个板块。

本教材适用于作为制药工程、药物化学专业本科教学的实验用书，也可用于与专业基础理论课“药物合成反应”相匹配的实验课教学。可供全国高等医药院校中药学、中药资源学与开发、中药制药专业使用，也可供成人教育本科和专升本、高等职业技术学校相关专业使用和参考。

图书在版编目（CIP）数据

药物合成反应实验/翟鑫主编．—3版．—北京：中国医药科技出版社，2019.7
全国高等医药院校药学类实验教材
ISBN 978-7-5214-1283-3

Ⅰ.①药…　Ⅱ.①翟…　Ⅲ.①药物化学-有机合成-化学实验-医学院校-教材
Ⅳ.①TQ460.31-33

中国版本图书馆CIP数据核字（2019）第159303号

美术编辑　陈君杞
版式设计　郭小平

出版　**中国健康传媒集团**｜中国医药科技出版社
地址　北京市海淀区文慧园北路甲22号
邮编　100082
电话　发行：010-62227427　邮购：010-62236938
网址　www.cmstp.com
规格　787×1092mm 1/16
印张　5 1/4
字数　94千字
初版　2006年3月第1版
版次　2019年7月第3版
印次　2019年7月第1次印刷
印刷　三河市航远印刷有限公司
经销　全国各地新华书店
书号　ISBN 978-7-5214-1283-3
定价　19.00元

获取新书信息、投稿、为图书纠错，请扫码联系我们。

前　言

“药物合成反应实验”是与制药工程、药物化学专业的专业基础理论课“药物合成反应”相匹配的实验课。开设本课程的目的是使学生在掌握理论课程内容的基础上，能够系统、直观地掌握药物合成中常见的有机单元反应与特殊反应的基本原理和基本操作，为后续开设的“药物化学实验”“制药工艺学实验”奠定基础，并有助于学生深入理解和掌握理论课讲授的内容。

本教材是在2014年出版的《药物合成反应实验（第二版）》的基础上进行完善与优化，其思路是着眼于“药物合成反应”理论课教学中重点介绍的九大反应类型，针对性地安排对应的实验内容，以达到实验教学与理论课程相辅相成、相互促进的目的。

2014年出版的《药物合成反应实验（第二版）》共收录了34个实验，每个实验设定为6学时，涵盖了药物合成中常见的大多数单元反应及基本操作过程，基本上满足了教学的需要，在学生实验技能的培养方面发挥了一定的作用。第二版教材是在2007年出版的《药物合成反应实验》（双语实验教材）的基础上，针对存在的“缺乏综合性、设计性实验内容，且每个实验间的联系不紧密”的情况，将内容整合为基础性实验、综合性实验和设计性实验三个板块，并重点改进了前版教材含有“药味”的实验内容偏少、个别实验的操作过程及使用的仪器过时等问题。但在第二版教材中，反应类型的覆盖面不全，仍有一些重要的反应类型没有很好地体现，且存在部分反应类型重复较多的现象。

鉴于上述情况，本版教材对第二版教材的相关内容进行修订，使之更好地满足实验教学的需求。为培养学生的实验技能服务，在遵循第二版教材的结构框架，即基础性实验、综合性实验、设计性实验三个板块的基础上，本版教材对实验内容做了较大的调整：一是去除了4个重复较多的反应类型实验如硝化反应、氧化反应等，增加了新的反应类型，如烃化反应、重排反应实验，使本教材所涉及的反应类型趋于完整；二是基于反应机制，调整本教材的实验顺序，使之与理论课教材顺序基本相一致，增加全书的可读性；此外，纠正了原教材中的个别错误。本版教材共计23个（次）实验，其主要特色如下。

（1）基于沈阳药科大学多年的药物合成反应实验教学的经验，借鉴国内众多院校的教学实例，使实验内容更加贴近教学实际，适当简化实验操作难度，增加实验中体现出的反应原理类型和基本操作类型，启发学生对相关反应的原理、反应条件、操作过程及产品分离纯化过程的思考与讨论。

（2）搜集文献，引入最新成果，更新内容动态，与时俱进。增加药物生产的应用实例的比重，例如新增加了卡托普利、硝苯地平、布洛芬等重要药物品种的合成应用实例。

（3）保留了开放性实验内容，即“吲哚美辛合成工艺”，通过指导学生独立完成吲哚美辛制备的各个环节，并得到吲哚美辛，要求学生开展文献调研、路线设计、合成及产物鉴定等步骤。

我们的期望是通过实验课的教学环节使学生掌握药物合成的基本原理、理论、方法及其在药物合成中的应用；进一步加深和巩固有机化学实验的操作技术以及相关的理论知识；培养学生理论联系实际的作风，事实求是、严格认真的科学态度和良好的工作习惯。

本实验教材是沈阳药科大学药物化学教研室全体同仁集体智慧的结晶，是我校在多年的实验课教学过程中所积累的宝贵经验的集体总结，在此要特别感谢前两版教材的所有编者，同时也要特别感谢本教材中所引用的实验操作原文的作者。

由于编写时间仓促，加之自身的水平有限，教材中难免存在疏漏与不足之处，恳请使用者提出宝贵意见，以便我们更好地修订完善。

编　者
2019 年 4 月

contents

目 录

第一部分　基础性实验 …… 1

实验一　氯代叔丁烷的制备 …… 1

Experiment 1　The preparation of *t* – butyl chloride …… 1

实验二　2,4 – 二氯乙酰苯胺的制备 …… 2

Experiment 2　The preparation of 2,4 – dichloroacetanilide …… 3

实验三　4 – 溴代正丁酸乙酯的制备 …… 4

Experiment 3　The preparation of ethyl 4 – bromo – 1 – butyrate …… 5

实验四　丙酰氯的制备 …… 6

Experiment 4　The preparation of Propionyl chloride …… 6

实验五　4,4′ – 二硝基二苯硫醚的制备 …… 7

Experiment 5　The preparation of 4,4′ – dinitrodiphenylsulfide …… 8

实验六　乙酰苯胺的制备 …… 8

Experiment 6　The preparation of Acetanilide …… 9

实验七　乙酰水杨酸的制备 …… 10

Experiment 7　The preparation of Acetylsalicylic acid …… 11

实验八　苯丙酮的制备 …… 12

Experiment 8　The preparation of Propiophenone …… 13

实验九　*N* – (4 – 甲基苯磺酰基) – 2 – 氨基苯甲酸甲酯的制备 …… 14

Experiment 9　The preparation of methyl – *N* – (4 – methylphenylsulfonyl) – 2 – amino benzoate …… 14

实验十　呋喃丙烯酸的制备 …… 15

Experiment 10　The preparation of Furanacrylic acid …… 16

实验十一　苯亚甲基苯乙酮（查尔酮）的制备 …… 17

Experiment 11　The preparation of Benzalacetophenone (Chalcone) …… 17

实验十二　硝苯地平的制备 …… 18

Experiment 12　The preparation of Nifedipine …… 19

实验十三　普萘洛尔的制备 …… 20

Experiment 13　The preparation of Propranolol …… 21

实验十四　2 – 巯基 – 4 – 甲基 – 6 – 羟基嘧啶的制备 …… 21

Experiment 14　The preparation of 2 – mercapto – 4 – methyl – 6 – hydropyrimidine …… 22

实验十五　4,4′-二硝基二苯砜的制备 …… 23
Experiment 15　The preparation of 4,4′-dinitrodiphenylsulfone …… 24
实验十六　对硝基苯甲醛的制备 …… 25
Experiment 16　The preparation of *p*-Nitrobenzaldehyde …… 26
实验十七　二苯甲烷的制备 …… 27
Experiment 17　The preparation of diphenylmethane …… 28
实验十八　二苯甲醇的制备 …… 29
Experiment 18　The preparation of Diphenyl carbinol …… 30
实验十九　对硝基乙酰苯胺的制备 …… 30
Experiment 19　The preparation of *p*-nitroacetanilide …… 31
第二部分　综合性实验 …… 33
实验二十　苯佐卡因的制备 …… 33
Experiment 20　The preparation of Benzocaine …… 36
实验二十一　L-(+)-扁桃酸乙酯的制备 …… 41
Experiment 21　The preparation of L-(+)-ethyl mandelate …… 45
实验二十二　布洛芬的制备 …… 49
Experiment 22　The preparation of Ibuprofen …… 52
第三部分　设计性实验 …… 57
实验二十三　吲哚美辛的合成及结构确证 …… 57
Experiment 23　The synthesis of Indometacin and its structural verification …… 61
附录 …… 68
参考文献 …… 75

第一部分　基础性实验

实验一　氯代叔丁烷的制备

【实验目的】

1. 学习叔醇的卤代反应机理、卤化剂的种类及特点。
2. 学习萃取、常压蒸馏操作。

【反应式】

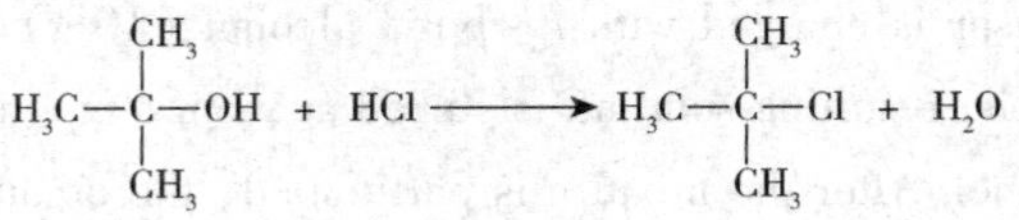

$$H_3C-\underset{CH_3}{\overset{CH_3}{C}}-OH + HCl \longrightarrow H_3C-\underset{CH_3}{\overset{CH_3}{C}}-Cl + H_2O$$

【实验试剂】

叔丁醇：10 g
浓盐酸：33 ml
碳酸氢钠（5 %）：15 ml

【实验操作】

在250 ml茄形瓶中，加入叔丁醇10 g、浓盐酸33 ml，加入搅拌子，安装回流冷凝器，室温下搅拌反应1 h。将反应液转移至分液漏斗中静置分层，分出有机层并依次用水（25 ml）、5 %碳酸氢钠溶液（15 ml）和水（25 ml）各洗涤1次，有机相以无水硫酸镁干燥30 min。常压蒸馏，收集50 ~ 53 ℃的馏分，得产品为无色透明液体，称重，计算收率。

【注意事项】

1. 5%碳酸氢钠溶液洗涤时注意排气。
2. 干燥剂的用量以保证容器中有可自由流动的干燥剂为宜。

【思考题】

1. 本实验中采用5 %碳酸氢钠洗涤的目的是什么？
2. 是否可以采用其他氯化剂？

Experiment 1　The preparation of *t* – butyl chloride

【Aim】

1. To comprehend the characteristics of chlorination reaction and chlorination reagents.

2. To learn the operation of extraction and distillation.

【Reaction equation】

$$H_3C-\underset{CH_3}{\overset{CH_3}{\underset{|}{\overset{|}{C}}}}-OH + HCl \longrightarrow H_3C-\underset{CH_3}{\overset{CH_3}{\underset{|}{\overset{|}{C}}}}-Cl + H_2O$$

【Reagents】

t – Butyl alcohol：10 g

Concentrated hydrochloric acid：33 ml

Sodium hydrogen carbonate（5 %）：15 ml

【Procedure】

A 250ml, three – necked, round – bottomed flask, fitted with a stirring bar, a thermometer, and a reflux condenser is charged with *t* – butyl alcohol（10 g）and concentrated hydrochloric acid（33 ml）. The resulting solution is stirred at room temperature for 1 h, then transfered to a separating funnel. After the mixture is partitioned, the organic layer is separated and washed with water（25 ml）, 5 % sodium hydrogen carbonate solution（15 ml）and water（25 ml）successively, then dried over anhydrous calcium chloride.

After removal of the solids by filtration through a Büchner funnel, the filtrate is subjected to a distillation. With the temperature at about 50 ~ 53 °C, the main fraction is collected as a colorless transparent liquid. The yiled can be calculated based on the weight of the product.

【Note】

1. Remember to release gas from the separating funnel after washing the organic layer with 5 % sodium hydrogen carbonate solution.

2. The amount of desiccant is suitable to ensure that there are free – flowing desiccants in the container.

【Subjects for Thinking】

1. What is the reason for washing the organic layer with 5 % sodium hydrogen carbonate solution?

2. What other kind of chlorinating agents can be used in this preparation?

实验二　2,4 – 二氯乙酰苯胺的制备

【实验目的】

1. 学习氯化反应的机理、氯化剂的种类及其特点。

2. 学习 HCl 气体吸收装置的应用。

【反应式】

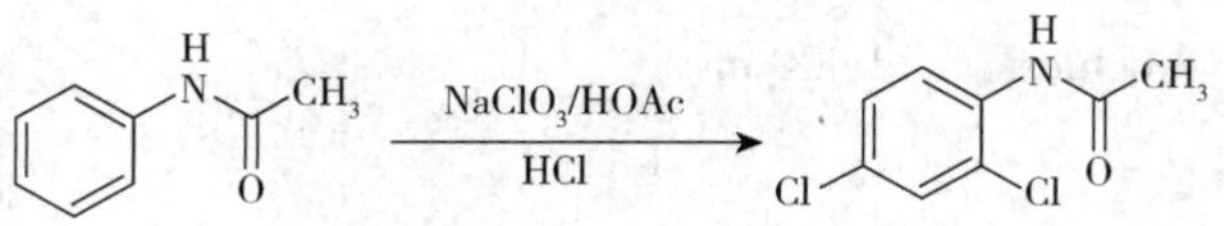

$$\text{C}_6\text{H}_5\text{NHCOCH}_3 \xrightarrow[\text{HCl}]{\text{NaClO}_3/\text{HOAc}} 2,4\text{-Cl}_2\text{C}_6\text{H}_3\text{NHCOCH}_3$$

【实验试剂】

乙酰苯胺：5 g

氯酸钠/水：3. 9 g/15 ml

冰醋酸：20 ml

浓盐酸：20 ml

【实验操作】

在 250 ml 的三颈瓶上配置搅拌子、温度计、回流冷凝器、滴液漏斗及气体吸收装置。将乙酰苯胺 5 g，冰醋酸 20 ml 加入反应瓶中，搅拌使之混合均匀，再加入浓盐酸 20 ml。将 3. 9 g 氯酸钠溶于 15 ml 水制得氯酸钠溶液，反应物在冰水浴冷却下滴加制备的氯酸钠溶液，控制滴加速度，使反应温度保持在 20 ~ 35 ℃，滴加完毕，于室温下继续搅拌反应 1. 5 h。抽滤，水洗滤饼至洗液呈中性，得 2，4 – 二氯乙酰苯胺粗品。以 80 % （V/V） 的甲醇重结晶得精品。

【注意事项】

1. 控制氯酸钠滴加速度（约 1 滴/10 s） 和反应温度，避免生成氯气速度过快。
2. 实验应严格在通风橱内进行。

【思考题】

1. 简述本实验的氯化原理。
2. 还可以选择哪些氯化剂完成此反应？

Experiment 2　The preparation of 2,4 – dichloroacetanilide

【Aim】

1. To comprehend the mechanism of chlorination reaction and the characteristics chloronation reagents.

2. To learn the application of HCl gas absorption device.

【Reaction equation】

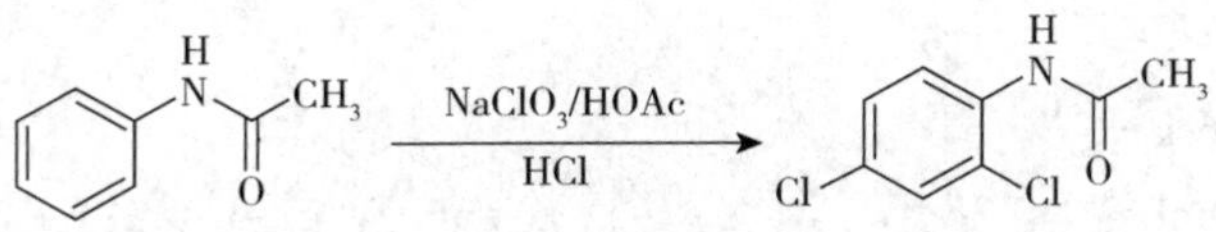

$$\text{C}_6\text{H}_5\text{NHCOCH}_3 \xrightarrow[\text{HCl}]{\text{NaClO}_3/\text{HOAc}} 2,4\text{-Cl}_2\text{C}_6\text{H}_3\text{NHCOCH}_3$$

【Reagents】

Acetanilide：5 g

Sodium chlorate/Water：3. 9 g/15 ml

Glacial acetic acid：20 mL

Concentrated hydrochloric acid：20 ml

【Procedure】

A 250 ml, three – necked flask equipped with a stiring bar, a thermometer, a reflux condenser and a dropping funnel is charged with 5 g of acetanilide, 20 ml of glacial acetic acid and 20 ml of concentrated hydrochloric acid. The reaction mixture is stirred in an icebath. A solution of 3. 9 g of sodium chlorate in 15 ml of water is added into the reaction mixture dropwise through a dropping funnel at such a rate so as to maintain the internal temperature between 20 ℃ and 35 ℃. The mixture is allowed stirring for 1. 5 h at room temperature. The crude 2, 4 – dichloroacetanilide is obtained after the filtration with a Büchner funnelfollowed by washing with cold – water washing. The fine product can be obtained by a recrystallization from 80 % (V/V) methanol.

【Note】

1. The addition speed of sodium chlorate (about 1 drop/10 s) and the internal temperature should be carefully controlled to avoid generating chlorine too fast.

2. This experiment should be carried out in a fume hood for safety.

【Subjects for Thinking】

1. What is the mechanism of chlorination reaction in this procedure?

2. Besides sodium chlorate, what kind of chlorination reagents can be used in this procedure?

实验三　4 – 溴代正丁酸乙酯的制备

【实验目的】

1. 学习内酯环开环卤代的反应机理。

2. 学习溴化氢气体的制备方法。

【反应式】

$$\gamma\text{-butyrolactone} \xrightarrow{C_2H_5OH/HBr} Br(CH_2)_3COOC_2H_5$$

【实验试剂】

γ – 丁内酯：26 g

无水乙醇：50 ml

【操作步骤】

在装有冷凝管和 HBr 气体吸收装置的 250 ml 三颈瓶中，加入无水乙醇 50 ml 和γ –

丁内酯26 g，冰水浴冷却至0 ℃，持续通入HBr气体1 h。将反应溶液置于0 ℃下放置2 h，将其倒入冰水中。用乙酸乙酯萃取水相（25 ml×3），合并有机相，无水硫酸镁干燥。过滤出干燥剂，浓缩滤液，浓缩液继续减压蒸馏，收集97～99 ℃/25 mmHg的馏分，得无色油状液体45 g，收率78 %。

溴化氢气体的制备：100 ml三颈瓶中加入9 ml四氢化萘，搅拌下缓慢滴加13 ml溴素，生成的气体经过四氢化萘洗气瓶洗涤后可直接使用。

【思考题】

本实验中是否可以选用其他溴化剂?

Experiment 3　The preparation of ethyl 4－bromo－1－butyrate

【Aim】

1. To comprehend the mechanism of halogenation reaction of lactone.

2. To learn the preparation method of hydrogen bromide.

【Reaction equation】

C_2H_5OH/HBr Br O OC_2H_5 O O

【Reagents】

γ－Butyrolactone：26 g

Anhydrous alcohol：50 ml

【Procedure】

A 250 ml, three－neck flask equipped with a reflux condenser connnecting with a HBr gas absorbent equipment and a set of HBr gas inlet tube, is charged with 26 g of γ－butyrolactone and 50 ml of anhydrous alcohol. The resulting mixture is then cooled to 0 ℃ in an ice－water bath. HBr gas is introduced, passing into the mixture uninterruptedly for 1 h. The resulting solution is allowed standing at 0 ℃ for additional 2 h, then carefully poured into ice water (50 g) The aqueous layer is extracted with ethyl acetate (25 ml×3), and the combined organic extractsare dried over anhydrous magnesium sulfate. After the desiccant is filtered off, the solvent is evaporated under reduced vaccum, and the residue is subjected to vacuum distillation. The target product is collected (96～99 ℃/25 mmHg) as a colorless oil with 78 % yield.

The preparation of HBr gas：A 100 ml, three－necked flask equipped with a stirring bar, a reflux condenser, a dropping funnel and a set of gas outlet tube, is charge with 9 ml of tetraline. 13 ml of bromine is added dropwise through the dropping funnel, and the HBr gas generated is ready to use after passing through a gas bottle containing tetraline.

【Subjects for Thinking】

What other bromination reagents can be used in this process?

实验四 丙酰氯的制备

【实验目的】

1. 学习羧酸氯化制备酰氯的机理。
2. 掌握氯化剂的种类及其特点。

【反应式】

$$3\ CH_3CH_2COOH + PCl_3 \longrightarrow 3\ CH_3CH_2COCl + H_3PO_3$$

【实验试剂】

丙酸：18.5 ml

三氯化磷：8.5 ml

【操作步骤】

在配置有回流冷凝器（顶端配有无水氯化钙干燥管及连接氯化氢气体吸收装置）的干燥 100 mL 圆底烧瓶中，加入丙酸 18.5 ml、三氯化磷 8.5 ml，在油浴上加热至 50 ℃，保温反应 3 h。冷却到室温后进行常压蒸馏，收集 76 ~ 80 ℃的馏分，得为无色透明液体产品，称重，计算收率。

【注意事项】

1. 反应开始阶段激烈放热，因此要注意控制反应温度。
2. 在气体吸收装置中可观察到有氯化氢气体放出。

【思考题】

1. 本实验可否选用其他氯化剂？
2. 简述蒸馏过程中馏分的取舍原则。

Experiment 4 The preparation of Propionyl chloride

【Aim】

1. To comprehend the chlorination mechanism of acyl chloride from acid.
2. To understand the characteristics of chloronation reagents.

【Reaction equation】

$$3\ CH_3CH_2COOH + PCl_3 \longrightarrow 3\ CH_3CH_2COCl + H_3PO_3$$

【Reagents】

Propionic acid：18.5 ml

Phosphorus trichloride：8. 5 ml

【Procedure】

A 100 ml, pre – dried round – bottomed flask fitted with a reflux condenser and a system for absorbing hydrogen chloride, is charged with 18. 5 ml of propionic acid and 8. 5 ml of phosphorus trichloride. The mixture is heated to 50 ℃ in an oil – bath for 3 h. The target product is collected (76 ~ 80 ℃) as a colorless liquid by distillation. The yiled can be calculated based on the weight of the product.

【Note】

1. This process is an exothermic reaction, so the heating speed should be controlled gentlely for safty.

2. The bubbles of hydrogen chloride gas can be observed in the gas – absorbing system.

【Subjects for Thinking】

1. What other chlorination reagents can be used in this procedure?

2. What is the principle of the distillation process?

实验五　4,4′ – 二硝基二苯硫醚的制备

【实验目的】

学习 *S* – 烃化反应的机理，烃化剂的种类及特点。

【反应式】

$$O_2N\text{-}C_6H_4\text{-}Cl + HS\text{-}C_6H_4\text{-}NO_2 \xrightarrow[H_2O, i\text{-}PrOH]{K_2CO_3} O_2N\text{-}C_6H_4\text{-}S\text{-}C_6H_4\text{-}NO_2$$

【实验试剂】

4 – 硝基苯硫酚：5. 0 g

4 – 硝基氯苯：5. 0 g

碳酸钾：4. 8 g

异丙醇：40 ml

【操作步骤】

在配置有搅拌子和回流冷凝器的 250 ml 圆底瓶中，依次加入 4 – 硝基苯硫酚5. 0 g、4 – 硝基氯苯 5. 0 g、碳酸钾 4. 8 g、水 50 ml 和异丙醇 40 ml，回流反应 6 ~ 8 h。反应结束后，冷却至室温，抽滤，滤饼用水/异丙醇（体积比 2：1）混合溶液洗涤 3 次（10 ml ×3），干燥得 4,4′ – 二硝基二苯硫醚粗品 8. 2 g，收率 95 %。

【思考题】

1. 本反应的反应机理如何？

2. 本反应若以氯苯为原料，是否可顺利进行？为什么？

Experiment 5　The preparation of 4,4′ – dinitrodiphenylsulfide

【Aim】

To comprehend the mechanism of *S* – alkylation, and the categories and characteristics of alkylation reagents.

【Reaction equation】

$$O_2N\text{-}C_6H_4\text{-}Cl + HS\text{-}C_6H_4\text{-}NO_2 \xrightarrow[H_2O,\,i\text{-}PrOH]{K_2CO_3} NO_2\text{-}C_6H_4\text{-}S\text{-}C_6H_4\text{-}NO_2$$

【Reagents】

4 – Nitrothiophenol：5.0 g

4 – Nitrochlorobenzene：5.0 g

Potassium carbonate：4.8 g

2 – Propanol：40 ml

【Procedure】

A 250 – ml, round – bottomed flask fitted with a stirring bar and a reflux condenser is charged with 4 – nitrothiophenol（5.0 g）, 4 – nitrochlorobenzene（5.0 g）, potassium carbonate（9.5 g）, 50 ml of water and 40 ml of *i* – propanol. The obtained mixture is refluxed for 6 to 8 hours. The flask is cooled to room temperature, and the precipitate is filtered off and washed with a solution of water/isopropanol（V/V = 2∶1）for three times（10 ml ×3）. The resulting solid is dried to give crude 4,4′ – dinitrodiphenylsulfide（8.2 g, 95 % yield）.

【Subjects for Thinking】

1. What is the reaction mechanism of this reaction?

2. Could this reaction succeed if chlorobenzene was used as the original material? Why?

实验六　乙酰苯胺的制备

【实验目的】

学习芳胺的酰化反应机理、酰化剂的种类及特点。

【反应式】

$$C_6H_5\text{-}NH_2 + H_3C\text{-}C(=O)\text{-}O\text{-}C(=O)\text{-}CH_3 \xrightarrow{H_2O} C_6H_5\text{-}NHCOCH_3$$

【实验试剂】

乙酸酐：7 ml

苯胺：5 ml

水：30 ml

【实验操作】

在配置好搅拌子、温度计、回流冷凝器及滴液漏斗的250 ml的三颈瓶中，加入5 ml苯胺及30 ml水，在搅拌下滴加7 ml乙酸酐，控制滴加速度以保证反应温度不超过40 ℃，滴加完毕于室温继续搅拌1 h，停止搅拌，室温下放置30 min。抽滤，以冷水洗涤滤饼至洗水呈中性，抽干，得乙酰苯胺粗品，以水为溶剂进行重结晶，可制得乙酰苯胺精品，干燥后测熔点、称重，计算收率。

【注意事项】

滴加乙酸酐速度不能过快，防止剧烈放热，使苯胺氧化，使产品变黄或变红。

【思考题】

1. 为什么要将粗产品用冷水洗至中性？
2. 本实验是否可选用其他酰化试剂？

Experiment 6 The preparation of Acetanilide

【Aim】

To comprehend the mechanism of acylation reaction, the types and characteristics of acylation reagents.

【Reaction equation】

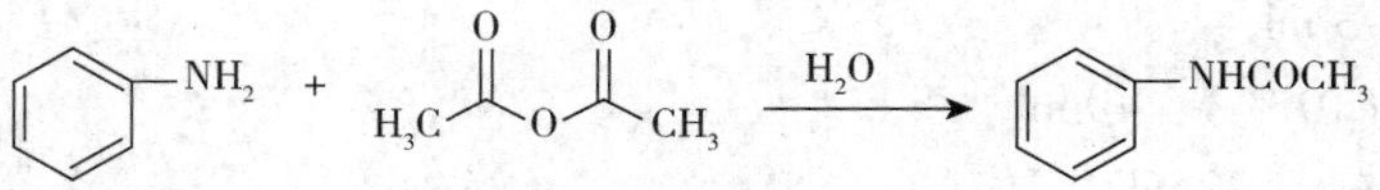

$$C_6H_5NH_2 + H_3C\text{-}CO\text{-}O\text{-}CO\text{-}CH_3 \xrightarrow{H_2O} C_6H_5NHCOCH_3$$

【Reagents】

Acetic anhydride: 7 ml

Aniline: 5 ml

Water: 30 ml

【Procedure】

A 250 - mL, three - necked flask equipped with a stirring bar, a thermometer, a reflux condenser and a dropping funnel, is charged with 5 mL of aniline and 30 mL of water. Then 7 mL of acetic anhydride is added into the reaction mixture dropwise through the dropping funnel at such a rate so as to maintain the internal temperature below 40 ℃. After complete addition, the reaction mixture is allowed stirring for 1 h at room temperature and standing for additional

30 min. The crude product was collected by filtration and washed with cold water. The fine acetanilide can be obtained by recrystallization from water. The yiled can be calculated based on the weight of the product.

【Note】

This exothermic reaction can easily cause the oxidization of aniline and turn the color of final product yellow or red. Therefore, the addition rate of acetic anhydride should not be too fast.

【Subjects for Thinking】

1. What is the reason for washing the crude product to neutral by cold water?

2. What other acylation reagents can be used in this preparation?

实验七　乙酰水杨酸的制备

【实验目的】

学习酚羟基的酰化反应机理、酰化剂的种类及特点。

【反应式】

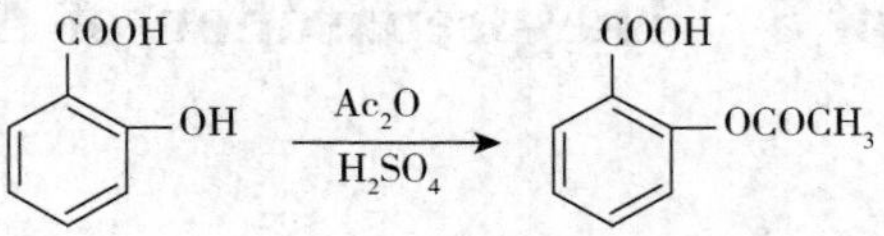

COOH, OH (水杨酸) $\xrightarrow[H_2SO_4]{Ac_2O}$ COOH, $OCOCH_3$ (乙酰水杨酸)

【实验试剂】

水杨酸：2.76 g

乙酸酐：8 ml

浓硫酸：0.3 ml

碳酸氢钠（10 %）：40 ml

盐酸（18 %）：20 ml

【实验操作】

在配有搅拌子、温度计和回流冷凝器的 250 ml 的三颈瓶中，分别加入水杨酸 2.76 g、乙酸酐 8 ml 和浓硫酸 0.3 ml，开动搅拌，水浴加热到 70 ℃反应 1 h，冷却到室温，向反应混合物中慢慢加入 15 ml 水，析出沉淀，抽滤，冷水洗涤滤饼，抽干得粗品。将粗品移至 250 ml 的烧杯中，搅拌下加入 10 % 碳酸氢钠 40 ml，搅拌至不再有气泡逸出，滤除不溶的副产物，将滤液慢慢倒入 20 ml 6 mol/L 的盐酸中，析出大量沉淀，待其冷却后抽滤，冰水洗涤滤饼，抽干得产品。以乙酸乙酯为溶剂进行重结晶可制得精制品，熔点 132 ~ 134 ℃。

【思考题】

1. 反应中加入浓硫酸的作用是什么？

2. 反应的主要的副产物是什么？怎样将其与产物分离？

3. 可用何种方法检测反应是否进行完全？

Experiment 7　The preparation of Acetylsalicylic acid

【Aim】

To comprehend the characteristics of acylation reaction and acylation reagents on hydroxyl group of phenols.

【Reaction equation】

COOH, OH (salicylic acid) $\xrightarrow[H_2SO_4]{Ac_2O}$ COOH, $OCOCH_3$ (acetylsalicylic acid)

【Reagents】

Salicylic acid：2. 76 g

Acetic anhydride：8 ml

Concentrated sulfuric acid：0. 3 ml

Sodium hydrogen carbonate（10%）：40 ml

Hydrochloric acid（18 %）：20 ml

【Procedure】

A 250 - ml, three - necked flask equipped with a stirring bar, a thermometer, a dropping funnel and a reflux condenser, is charged with 2. 76 g of salicylic acid, 8 ml of acetic anhydride and 0. 3 ml of concentrated sulfuric acid. The resulting mixture is stirred at 70 ℃ for 1 h, then cooled to room temperature, at which point 15 ml of cold water is added slowly. The preciptate is collected by filtration, washed with water, then transferred into a 250 ml beaker. 40 ml of sodium hydrogen carbonate solution（10 %）is added in and the mixture is allowed stirring well till no more CO_2 gas is released. The insoluble by - product is filtered off, and the filtrate is added into a beaker containing 20 ml of hydrochloric acid（18 %）to form a white precipitate, which is collected by filtration and washed with water. A recrystallization from ethyl acetate give the fine product as white solid. m. p 132 ~ 134 ℃.

【Subjects for Thinking】

1. What is the function of concentrated sulfuric acid in this procedure?

2. What is the side reaction in this preparation and how to avoid it?

3. Which method can be used to detect whether the reaction is complete?

实验八　苯丙酮的制备

【实验目的】

1. 学习 Friedel - Crafts 酰化反应的机理、反应条件以及操作过程。

2. 学习减压蒸馏的操作。

【反应式】

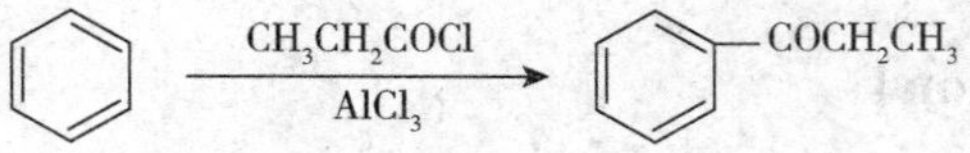

$$C_6H_6 \xrightarrow[AlCl_3]{CH_3CH_2COCl} C_6H_5COCH_2CH_3$$

【实验试剂】

丙酰氯：9.2 g

苯：35 mL

三氯化铝：14 g

氢氧化钠（5 %）：45 ml

【操作步骤】

在配置有搅拌子、温度计和回流冷凝器（顶端装有无水氯化钙干燥管及连接氯化氢气体吸收装置）的 250 ml 的三颈瓶中，加入三氯化铝 14 g，干燥过的苯 35 ml，搅拌下滴加丙酰氯 9.2 g，控制滴加速度使反应温度维持在 25 ~ 30 ℃为宜。滴加完毕，缓慢升温到 50 ℃，在此温度下继续反应 2 h，冷却至 20 ℃，将反应混合物倒入 60 g 碎冰中，用少量盐酸将析出的沉淀溶解，分出有机层，水层用苯提取 3 次（每次15 ml），合并有机相后再用 5 % 的氢氧化钠洗涤 3 次（每次 15 ml），最后再用水洗一次，以无水硫酸镁干燥。

将干燥剂滤除，安装好减压蒸馏装置，先常压蒸出苯，然后再减压蒸出产品，收集 92 ~95 ℃/11 mmHg 的馏分。

【注意事项】

1. 称取三氯化铝时动作要快，要在通风橱内进行，以防其吸潮分解。

2. 本实验中最好选用不含噻吩的苯。

3. 冰解一步要在通风橱内进行。

4. 事先要预习有关减压蒸馏的原理及操作。

【思考题】

1. 本实验中是否可以选用丙酸酐为酰化剂？

2. 反应的后处理过程中加入盐酸的作用是什么？

Experiment 8 The preparation of Propiophenone

【Aim】

1. To comprehend the mechanism, reaction conditions and operation of Friedel – Crafts reaction.

2. To master the operation of vacuum distillation.

【Reaction equation】

$$C_6H_6 \xrightarrow[AlCl_3]{CH_3CH_2COCl} C_6H_5COCH_2CH_3$$

【Reagents】

Propionyl chloride: 9.2 g

Benzene: 35 ml

Aluminium chloride: 14 g

Sodium hydroxide (5%): 45 ml

【Procedure】

A 250 ml, three – necked flask equipped with a stirring bar, a thermometer, a dropping funnel and a reflux condenser fitted with a calcium chloride – filled drying tube, is charged with 14 g of aluminium chloride and 35 ml of dry benzene. To the resulting mixture, 9.2 g of propionyl chloride is added dropwise through the dropping funnel at such a rate so as to maintain the internal temperatur at 25 ~ 30 ℃. After complete addition, the mixture is allowed stirring at 50 ℃ for 2 h. After cooled to 20 ℃, the reaction mixture is poured carefully into a beaker containing crushed ice (60 g). The resulting sediment is dissolved with diluted hydrochloric acid. The organic layer is separated, and the aqueous layer is extracted with benzene (15 ml × 3). The combined organic layer is washed with 5% sodium hydroxide solution (15 ml × 3), and water (15 ml), then dried over anhydrous magnesium sulfate.

After the desiccantis filtered off, the filtrate is subjected an atmosphere distillation to remove residual benzene and a vacumm distillation to afford propiophenone (92 ~ 95 ℃/11 mmHg) as a colorless liquid.

【Note】

1. This process should be operated rapidly in the fume hood to avoid moisture absorption.

2. It is recommended to use benzene that does not contain thiophene.

3. This operation should be conducted in the fume hood.

4. Please review the operation ofvacumm distillation in advance.

【Subjects for Thinking】

1. Can propanoic anhydride be used in this procedure?

2. What is the reason for the addition of hydrochloric acid in the work – up procedure?

实验九　*N*–（4–甲基苯磺酰基）–2–氨基苯甲酸甲酯的制备

【实验目的】

学习芳胺 *N*–磺酰化反应及酰化剂的特点。

【反应式】

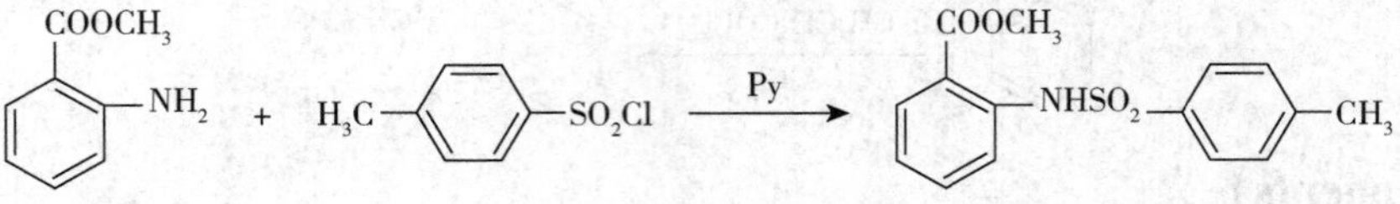

$$\text{2-}H_2N\text{-}C_6H_4\text{-}COOCH_3 + H_3C\text{-}C_6H_4\text{-}SO_2Cl \xrightarrow{Py} \text{2-}(CH_3\text{-}C_6H_4\text{-}SO_2NH)\text{-}C_6H_4\text{-}COOCH_3$$

【实验试剂】

对甲苯磺酰氯：10 g

2–氨基苯甲酸甲酯：5. 5 ml

吡啶：20 ml

【操作步骤】

在 250 ml 三颈瓶中加入对甲苯磺酰氯 10 g 和干燥吡啶 20 ml，搅拌下滴加邻氨基苯甲酸甲酯 5. 5 ml。滴毕，室温搅拌 3 h。滤出固体，用 20 ml 甲醇洗涤滤饼，干燥，将所得固体用甲醇重结晶，得到 10 g 白色晶体，收率 81%，m. p 110 ~ 113℃。

【思考题】

1. 此反应是否可以选用其他酰化剂？
2. 反应中吡啶的作用是什么？

Experiment 9　The preparation of methyl – *N* –（4 – methylphenylsulfonyl）– 2 – amino benzoate

【Aim】

To comprehend the characteristics of *N* – sulfonylation reaction and sulfonylation reagents.

【Reaction equation】

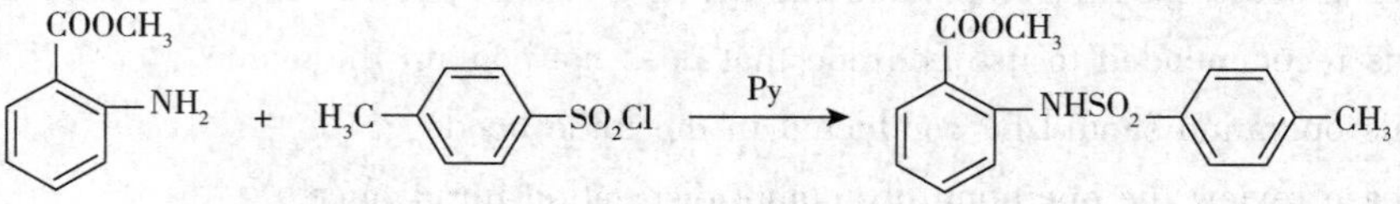

$$\text{2-}H_2N\text{-}C_6H_4\text{-}COOCH_3 + H_3C\text{-}C_6H_4\text{-}SO_2Cl \xrightarrow{Py} \text{2-}(CH_3\text{-}C_6H_4\text{-}SO_2NH)\text{-}C_6H_4\text{-}COOCH_3$$

【Reacgents】

p – Methylbenzensulfonyl chloride：10 g

Methyl 2 - aminobenzoate: 5. 5 ml

Pyridine: 20 ml

【Procedure】

A 250 ml, three - necked flask equipped with a stirring bar, a thermometer, a dropping funnel and a reflux condenser, is charged with 10 g of 4 - methylbenzensulfonyl chloride and 20 ml of dried pyridine is added. To this solution, 5. 5 mL of methyl 2 - aminobenzoate is added dropwise through the dropping funnel. After complete addition, the reaction mixture is allowed stirring at room temperature for 3 h. The crude product is separated by filtration and washed with 20 mL of methanol. A recrystallization from methanol give the fine product with a yield of 81 %. m. p 110 ~ 113 ℃.

【Subjects for Thinking】

1. What other sulfonylation reagents can be used in this prepartion?

2. What is the reason for the addition of pyridine in the process?

实验十 呋喃丙烯酸的制备

【实验目的】

学习 Knoevenagel 反应的机理、反应条件及特点。

【反应式】

$$\text{(furan-2-yl)}-CHO + CH_2(COOH)_2 \xrightarrow[\triangle]{Py} \text{(furan-2-yl)}-CH{=}CH-COOH$$

【实验试剂】

丙二酸：2. 6 g

呋喃甲醛：2. 4 g

吡啶：1. 2 mL

【操作步骤】

将 2. 4 g 呋喃甲醛、2. 6 g 丙二酸和 1. 2 ml 吡啶加入到 50 ml 的圆底烧瓶中，安装回流冷凝器，在 90 ℃水浴上加热反应 2 h。冷却到室温后，将反应液转移到 250 ml 的烧杯中，加入 50 ml 水稀释反应液，再加入浓氨水使反应物全部溶解，抽滤，少量水淋洗，合并滤液及洗液，在搅拌下以 18 % 盐酸调 pH = 3，待酸化液充分冷却后，抽滤，水洗滤饼两次（每次约 20 ml），干燥后得呋喃丙烯酸粗品，乙醇重结晶可得精品，熔点 139 ~ 140 ℃。

【注意事项】

1. 本实验所用的原料应事先进行干燥处理，呋喃甲醛应该是无色液体，但由于易被氧化，时间常后常呈黄色或棕色。

2. 氨水的用量约 10 ml。

【思考题】

1. 实验中吡啶的作用是什么？

2. 还可以采用其他何种反应同样可制备该化合物？

Experiment 10 The preparation of Furanacrylic acid

【Aim】

To comprehend the mechanism, characteristics and reaction conditions of Knoevenagel reaction.

【Reaction equation】

$$\text{(furan-2-yl)}\text{—CHO} + CH_2(COOH)_2 \xrightarrow[\triangle]{Py} \text{(furan-2-yl)}\text{—CH=CH—COOH}$$

【Reagents】

Malonic acid：2. 6 g

Furaldehyde：2. 4 g

Pyridine：1. 2 mL

【Procedure】

A 50 – ml, round – bottomed flask fitted with a reflux condenser is charged with 2. 4 g of fresh distilled furaldehyde, 2. 6 g of malonic acid and 1. 2 ml of dry pyridine. The resulting mixture is heated on a 90 ℃ water bath for 2 h, then cooled to room temperature and transferred into a 250 – ml beaker, and then diluted with 50 ml of water. Concentrated aqueous ammonia is added to dissolve the solid, and the mixture is filtered and washed with a little water. The combined filtrate is charged with 18 % hydrochloric acid to adjust its pH value to 3. After cooled to room temperature, the crude product is collected by filtration and washed by water (20 ml × 2). A recrystallization from ethanol give the fine product. m. p 139 ~ 140 ℃.

【Note】

1. It is essential to dry all the starting materials before use.

2. 10 mL of aqueous ammoniais used in the adjustion process.

【Subjects for Thinking】

1. What is the function of pyridine in this procedure?

2. Are there other reactions can also be applied to generate the title compound?

实验十一 苯亚甲基苯乙酮（查尔酮）的制备

【实验目的】

学习 Aldol 缩合反应的机理、特点及反应条件。

【反应式】

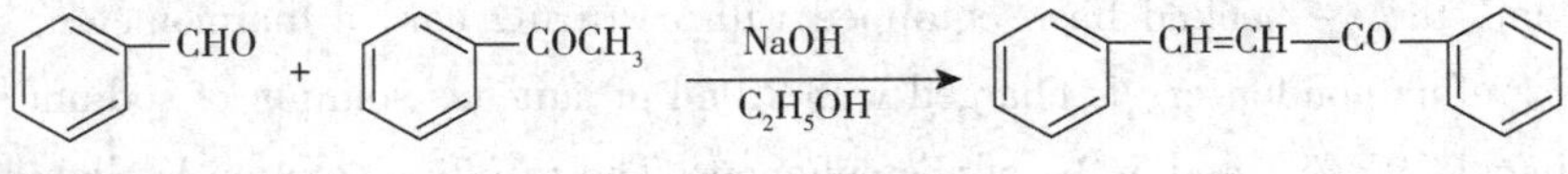

$$C_6H_5CHO + C_6H_5COCH_3 \xrightarrow[C_2H_5OH]{NaOH} C_6H_5CH{=}CH{-}CO{-}C_6H_5$$

【实验试剂】

苯甲醛：4.6 g

苯乙酮：5.2 g

乙醇（95 %）：15 ml

氢氧化钠/水：2.2 g /20 ml

【操作步骤】

在配有搅拌子、温度计、回流冷凝器及滴液漏斗的 250 ml 的三颈瓶中，加入氢氧化钠水溶液 20 ml、95 % 乙醇 15 ml 及苯乙酮 5.2 g，水浴加热到 20 ℃，滴加苯甲醛 4.6 g，滴加过程中维持反应温度 20 ~ 25 ℃。加毕，于该温度下继续搅拌反应 0.5 h。加入少量的查尔酮做晶种，继续搅拌 1.5 h，析出沉淀。抽滤，水洗至洗液呈中性，抽干得查尔酮粗品。以乙酸乙酯为溶剂重结晶，得精品为浅黄色针状结晶，熔点 55 ~ 56 ℃。

【注意事项】

1. 反应温度不要超过 30 ℃或低于 15 ℃，温度过高副产物增多；温度过低产物黏稠，不易过滤。

2. 加入晶种后，一般搅拌 1.5 h 后即可析出产品结晶。

【思考题】

1. 本实验中可能的副反应有哪些？怎样可以避免？

2. 为什么该产品析晶较困难？

Experiment 11 The preparation of Benzalacetophenone（Chalcone）

【Aim】

To comprehend the mechanism, characteristics and reaction conditions of Aldol reaction.

【Reaction equation】

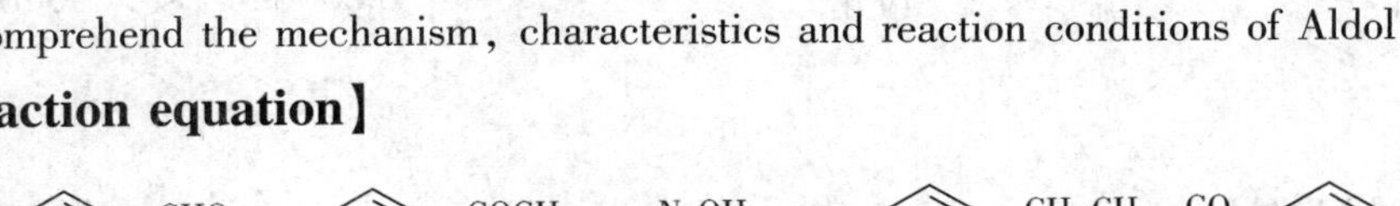

$$C_6H_5CHO + C_6H_5COCH_3 \xrightarrow[C_2H_5OH]{NaOH} C_6H_5CH{=}CH{-}CO{-}C_6H_5$$

【Reagents】

Benzaldehyde：4. 6 g

Acetophenone：5. 2 g

Ethanol（95 %）：15 ml

Sodium hydroxide/H_2O：2. 2 g/20 ml

【Procedure】

A 250 ml，three－necked flask equipped with a stirring bar，a thermometer，a dropping funnel and a reflux condenser，is charged with 20 ml of aqueous solution of sodium hydroxide，15 ml of ethanol（95 %）and 5. 2 g of acetophenone. The resulting solution is slightly heated to 20℃，and 4. 6 g of benzaldehyde is added dropwise at such a rate so as to maintain the internal temperature at 20～25 ℃. After compelte addition，the mixture is allowed stirring for 0. 5 h，at which point a small amount of powdered benzalacetophenone is added as crystal seed. The mixture is stirred for an additional 1. 5 h. The crude product is collected by filtration and washed with water. A recrystallization from ethyl acetate gives the fine product. m. p 55～56 ℃.

【Note】

1. Suitable temperature for this reaction is between 15 and 30 ℃. Higher temperature will lead to the increase of the byproducts，while lower temperature will make the product too thick and hard to be filtered.

2. After adding the crystal seeds，the product will separate crystals after 1. 5 h of stirring.

【Subjects for Thinking】

1. What are the possible side reactions in this preparation? How to avoid them?

2. Why is it difficult for this product to crystallize?

实验十二　硝苯地平的制备

【实验目的】

1. 掌握 Hantzsch 吡啶合成法的反应机理和实验操作。

2. 了解多组分反应的概念、特点和优势。

【反应式】

CHO, NO_2 + H_3C–CO–CH_2–CO–OCH_3 + NH_4HCO_3 $\xrightarrow{CH_3OH}$ NO_2, H_3COOC, $COOCH_3$, H_3C, N, H, CH_3

【实验试剂】

2－硝基苯甲醛：10. 0 g

乙酰乙酸甲酯：20. 0 ml
碳酸氢铵：8. 5 g
甲醇：30. 0 ml

【操作步骤】

在配有搅拌子、温度计和回流冷凝管的 250 mL 三颈瓶中，加入乙酰乙酸甲酯20. 0 ml和甲醇30. 0 ml。室温搅拌下，加入邻硝基苯甲醛 10. 0 g，加毕，升温至 50 ℃，搅拌反应 30 min。分批加入碳酸氢铵 8. 5 g，搅拌 15 min 后加热至回流，继续反应 3 h。反应完毕，冷却到室温，析出沉淀。抽滤，用少量甲醇洗涤滤饼，压干得 17. 5 g 硝苯地平粗产品，收率 76. 4 %。

【思考题】

1. 本反应的反应机理如何？
2. 在本反应中，除了碳酸氢铵外，还有哪些物质可以作为氨的来源？
3. 多组分反应有哪些优势？

Experiment 12 The preparation of Nifedipine

【Aim】

1. To learn the mechanism and operation method of Hantzsch pyridine reaction.
2. To understand the concept, application and advantages of multi – component reaction.

【Reaction equation】

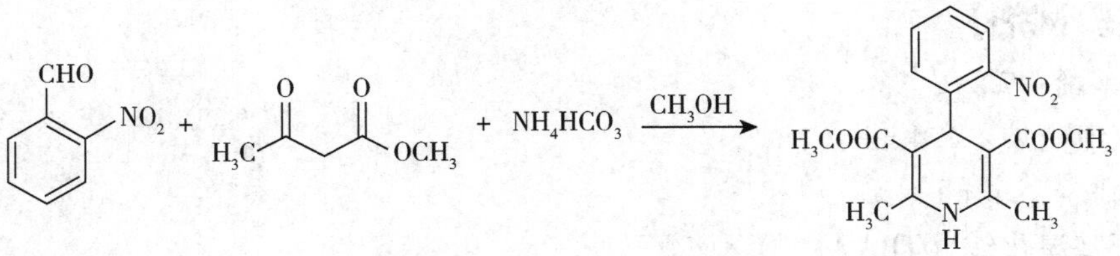

【Reagents】

2 – Nitobenzaldehyde：10. 0 g
Methyl acetoacetate：20. 0 ml
Ammonium bicarbonate：8. 5 g
Methanol：30. 0 ml

【Procedure】

A 250 ml, three – necked flask equipped with a stirring bar, a thermometer and a reflux condenser, is charged with 20. 0 mL of methyl acetoacetate and 30. 0 mL of methanol. To this solution, 10. 0 g of 2 – nitobenzaldehyde is added under stirring at room temperature. The resulting mixture is heated to 50 ℃ and stirred for 30 min. Then 8. 5 g of ammonium bicarbonate is added gradually. The mixture is allowed stirring for 15 min at 50 ℃ and for 3 h under reflux, and then cooled to room temperature, resulting in a quantity of precipitate. The crude product is collected by filtration and wahed with methanol. The final product (17. 5 g, 76. 4 % yield) is obtained after drying treatment.

【Subjects for Thinking】

1. What is the mechanism of this reaction?

2. What other substances can be used as the source of ammonia in addition to ammonium bicarbonate?

3. What are the advantages of multi - component reaction?

实验十三　普萘洛尔的制备

【实验目的】

1. 掌握环氧烷类烃化剂的特点及反应条件。

2. 了解相转移催化剂的种类、特点及反应条件。

【反应式】

OH　Cl　O　+　+　CH_3　H_2N　CH_3　TBAB　CH_3　O　N　CH_3　H　OH

【实验试剂】

α-萘酚：7. 2 g

环氧氯丙烷：4. 4 ml

异丙胺：6. 4 ml

四丁基溴化铵（TBAB）：0. 8 g

【操作步骤】

在配有搅拌子、温度计、回流冷凝器的250 ml三颈瓶中，分别加入α-萘酚7. 2 g、环氧氯丙烷4. 4 ml、异丙胺6. 4 ml以及四丁基溴化铵0. 8 g。冰水浴冷却下搅拌10 min后，缓慢滴加10 %氢氧化钠溶液30 ml。滴加完毕，于0 ℃继续搅拌1. 5 h，撤去冰水浴，升温至100 ℃反应1 h。停止加热，用冰水浴冷却至0 ℃，滴加10 %氢氧化钠溶液30 ml，有沉淀析出。抽滤，少量水洗滤饼，压干得普萘洛尔粗产品11. 0 g，收率84. 9 %。

【思考题】

1. 本反应中可能产生的副产物有哪些？

2. 相转移催化剂TBAB在反应中起到什么作用？

Experiment 13 The preparation of Propranolol

【Aim】

1. To comprehend the characteristics and reaction conditions of epoxide alkylation agent.

2. To understand the categories, characteristics and reaction conditions of phase transfer catalyst.

【Reaction equation】

OH, Cl, O, + , + , H_2N, CH_3, CH_3, TBAB, O, N, H, CH_3, CH_3, OH

【Reagents】

α – Naphthol: 7. 2 g

Epichlorohydrin: 4. 4 ml

Isopropylamine: 6. 4 ml

Tetrabutylammonium bromide (TBAB): 0. 8 g

【Procedure】

A 250 – ml, three – necked flask equipped with a stirring bar, a thermometer and a reflux condenser, is charged with 7. 2 g of α – naphthol, 4. 4 mL of epichlorohydrin, 6. 4 ml of isopropylamineand and 0. 8 g of TBAB. The mixture is cooled to 0 °C under an ice – water bath, then 30 ml of 10 % sodium hydroxide solution is added slowly. After complete addition, the mixture is allowed stirring for 1. 5 h at 0 ℃ and for 1 h at 100 ℃, and then cooled back to 0 ℃. Then, 30 ml of 10 % sodium hydroxide solution is added dropwise to form a white precipitate. The crude product is collected by filtration and washed with water. The final product (11. 0 g, 84. 9 % yield) is obtained after drying treatment.

【Subjects for Thinking】

1. What are the possible side reactions in this preparation?

2. What is the role of phase transfer catalyst TABA in the process?

实验十四 2 – 巯基 – 4 – 甲基 – 6 – 羟基嘧啶的制备

【实验目的】

1. 学习环合反应的机理及反应条件。

2. 学习金属钠的使用方法及操作注意事项。

【反应式】

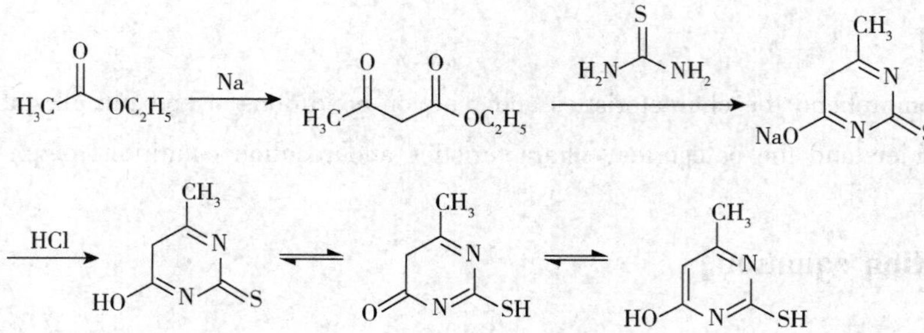

$$H_3C-CO-OC_2H_5 \xrightarrow{Na} H_3C-CO-CH_2-CO-OC_2H_5 \xrightarrow{H_2N-CS-NH_2} \text{(4-methyl-2-thioxo-pyrimidin-6-olate sodium, NaO)} \xrightarrow{HCl} \text{(HO, =S form)} \rightleftharpoons \text{(O=, SH form)} \rightleftharpoons \text{(HO, SH form)}$$

【实验试剂】

乙酸乙酯：30 ml

金属钠：1. 4 g

硫脲：2. 8 g

【操作步骤】

在装有搅拌子、温度计、回流冷凝器及滴液漏斗的250 mL三颈瓶中，加入乙酸乙酯30 ml，开动搅拌，将1. 4 g金属钠切成小块分批加入到反应瓶中，反应放热，加毕，油浴加热至回流使金属钠全部溶解。加入硫脲2. 8 g，继续回流反应4 h，停止反应，向反应液中加入20 ml水，使反应物全部溶解，再加入0. 5 g活性炭脱色5 min，抽滤，滤液以10 %的盐酸酸化至pH =3，析出沉淀，抽滤，水洗滤饼至洗水呈中性，抽干得粗产品，乙醇重结晶可制得精品。

【注意事项】

1. 反应放热可能使反应回流。

2. 金属钠溶解后即刻析出乙酰乙酸乙酯的钠盐沉淀。

【思考题】

1. 环合反应一步的机理是怎样的?

2. 可否用醇钠代替反应中的金属钠?

Experiment 14　The preparation of 2 – mercapto – 4 – methyl – 6 – hydropyrimidine

【Aim】

1. To comprehend the mechanism and reaction conditions of cyclization reaction.

2. To comprehend the function of sodium and the operation attentions for it.

【Reaction equation】

$$H_3C-C(=O)-OC_2H_5 \xrightarrow{Na} H_3C-C(=O)-CH_2-C(=O)-OC_2H_5 \xrightarrow{H_2N-C(=S)-NH_2} \text{(sodium salt, NaO-, CH}_3\text{, =S pyrimidine)} \xrightarrow{HCl} \text{(HO-, CH}_3\text{, =S)} \rightleftharpoons \text{(O=, CH}_3\text{, SH)} \rightleftharpoons \text{(HO-, CH}_3\text{, SH)}$$

【Reagents】

Ethyl acetate: 30 ml

Sodium: 1. 4 g

Thiourea: 2. 8 g

【Procedure】

A 250 ml, three - necked flask equipped with a stirring bar, a thermometer, a dropping funnel and a reflux condenser, is charged with 30 ml ofethyl acetate and 1. 4 g of sodium (in small pieces). After complete addition, the mixture is refluxed till the sodium is fully dissolved. At this point, 2. 8 g of thiourea is added and the resulting mixture is refluxed for 4 h. 20 ml of warm water is then poured into the flask to dissolve the precipitate. The resulting solution is charged with 0. 5 g of activated carbon for 5 min and filtered. The filtrate is treated with 10 % hydrochloric acid to adjust pH = 3, resulting in a quantity of sediment. The product is collected by filtration and washed with water. A recrystallization from ethyl alcohol gives the fine product.

【Note】

1. This exothermic reaction may cause the solution to reflux.

2. The sodium salt of ethyl acetoacetate will precipitate soon after the fully dissolution of sodium pieces.

【Subjects for Thinking】

1. What is the mechanism of the cyclization reaction?

2. Can sodium alkoxide be used to replace the sodium in this preparation?

实验十五 4,4′-二硝基二苯砜的制备

【实验目的】

1. 学习砜的制备方法。

2. 学习过氧化氢为氧化剂的反应机理、条件和特点。

【反应式】

$$\text{4,4'-}(O_2N\text{-}C_6H_4)_2S \xrightarrow[H_2O_2,AcOH]{H_2O_2,Na_2WO_4} \text{4,4'-}(O_2N\text{-}C_6H_4)_2SO_2$$

【实验试剂】

4,4′-二硝基二苯硫醚：8.2 g

钨酸钠二水合物：0.2 g

30 %过氧化氢：7.0 ml

冰醋酸：40 ml

甲基异丁基酮：80 ml

【操作步骤】

在配有搅拌子、温度计和回流冷凝器的250 ml三颈瓶中，依次加入4,4′-二硝基二苯硫醚8.2 g，冰醋酸40 ml和含有0.20 g钨酸钠二水合物的水溶液0.7 ml。将反应液升温至75～80 ℃，缓慢滴加30%的过氧化氢溶液7.0 ml，控制滴加速度，使反应温度不超过90 ℃，大约15 min滴完，于75～80℃继续反应2.5 h。冷至室温，抽滤，滤饼用水洗涤三次（20 ml×3），干燥，得粗品8.3 g。

将粗品加入80 ml的甲基异丁基酮中，回流2 h。冷至室温，抽滤，干燥得精品8.0 g，收率87 %。熔点282 ℃。

【思考题】

1. 钨酸钠在该反应中有什么作用？

2. 可否选用其他的氧化剂？

Experiment 15 The preparation of 4,4′-dinitrodiphenylsulfone

【Aim】

1. To learn the preparation of sulfone derivatives.

2. To comprehend the mechanism of hydrogen peroxide oxidation, their characteristics and reaction conditions.

【Reaction equation】

$$\text{4,4'-}(O_2N\text{-}C_6H_4)_2S \xrightarrow[H_2O_2,AcOH]{H_2O_2,Na_2WO_4} \text{4,4'-}(O_2N\text{-}C_6H_4)_2SO_2$$

【Reagents】

4,4′-Dinitrodiphenylsulfide: 8.2 g

Sodium tungstate dihydrate：0. 2 g

30% Hydrogen peroxide：7. 0 ml

Glacial acetic acid：40 ml

Methylisobutylketone：80 ml

【Procedure】

A 250 – ml, three – necked flask equipped with a stirring bar, a thermometer and a reflux condenser, is charged with 8. 2 g of 4,4′ – dinitrodiphenylsulfide, 40 ml of glacial acetic acid, and 0. 7 ml of aqueous solution of sodium tungstate dihydrate（0. 2 g）. The resulting mixture is heated to 75 ~ 80 ℃, and 7. 0 ml of 30 % hydrogen peroxide is added dropwise within 15 min , at such a rate so as to maintain the internal temperature no higher than 90 °C. After complete addition, the mixture is allowed stirring for additional 2. 5 h at 75 ~ 80 ℃, and cooled to room temperature. The crude product is collected by filtration and washed with water（20 ml × 3）. A recrystallization from methylisobutylketone gives the fine product（8. 0 g, 87 % yield）. m. p 282 ℃.

【Subjects for Thinking】

1. What is the role of the sodium tungstate dehydrate in this reaction?

2. Can other oxidative reagents be used in this reaction?

实验十六　对硝基苯甲醛的制备

【实验目的】

学习苄位氧化成醛的常用氧化剂的种类、特点及反应条件。

【反应式】

CH_3–C_6H_4–NO_2 $\xrightarrow{CrO_3/Ac_2O}$ $CH(OAc)_2$–C_6H_4–NO_2 $\xrightarrow{H_2O/H_2SO_4}$ CHO–C_6H_4–NO_2

【实验试剂】

对硝基甲苯：6. 3 g

醋酐：50 ml + 57 ml

铬酸酐：12. 5 g

硫酸：10 ml + 2 ml

碳酸钠（2 %）：40 ml

【操作步骤】

在 250 mL 的三颈瓶上配置搅拌子、温度计、回流冷凝器及滴液漏斗，将醋酐50 ml

及对硝基甲苯6.3 g加入反应瓶中，冰盐浴冷却下加入浓硫酸10 ml，冷却到0 ℃，在搅拌下滴加事先制好的铬醋酐溶液，维持反应温度在10 ℃以下。加毕，于5～10 ℃反应2 h，将反应混合物倒入250 g碎冰中，搅拌均匀后再以冰水稀释至750 ml，抽滤，将滤饼悬浮于40 ml 2 %的碳酸钠溶液中，充分搅拌后抽滤，依次用水、乙醇洗涤滤饼，抽干后得对硝基苯甲醛二醋酸酯粗品。

将上述制得的对硝基苯甲醛二醋酸酯粗品置于100 ml的圆底瓶中，加入水20 ml、乙醇20 ml和浓硫酸2 ml，加热回流30 min，趁热抽滤，滤液在冰水中冷却后析出结晶，抽滤、水洗，干燥后得产品，m. p 106～107 ℃。

【注意事项】

1. 铬醋酐溶液配制：将铬酐在搅拌下分批加入到醋酐中，不能反加料，否则易爆炸。

2. 滤液用50 ml水稀释后还可析出部分产品。

【思考题】

1. 将芳环上的甲基氧化成醛时，为何需经过二醋酸酯一步？

2. 本实验除了可采用铬醋酐为氧化剂外，还可以选用哪些氧化剂？

Experiment 16　The preparation of *p* – Nitrobenzaldehyde

【Aim】

To comprehend the oxidant categories, characteristics and reaction conditions of oxidation of toluene to benzaldehyde.

【Reaction equation】

$$p\text{-}O_2N\text{-}C_6H_4\text{-}CH_3 \xrightarrow{CrO_3/Ac_2O} p\text{-}O_2N\text{-}C_6H_4\text{-}CH(OAc)_2 \xrightarrow{H_2O/H_2SO_4} p\text{-}O_2N\text{-}C_6H_4\text{-}CHO$$

【Reagents】

p – Nitrotoluene: 6.3 g

Acetic anhydride: 50 ml + 57 ml

Chromium trioxide: 12.5 g

Sulfuric acid: 10 ml + 2 ml

Sodium carbonate (2 %): 40 ml

【Procedure】

A 250 ml, three – necked flask equipped with a stirring bar, a thermometer, a dropping funnel and a reflux condenser, is charged with 50 ml of acetic anhydride, 6.3 g of *p* – nitrotoluene and 10 ml of concentrated sulfuric acid. The resulting mixture is cooled to 0 ℃ by an ice water bath, at which point a freshly prepared solution of 12. 5 g of chromium trioxide in 57 ml acetic anhydride is added dropwise at a such rate so as to maintain the internal temperature no higher than 10 °C. After complete addition, the resulting mixture is allowed stirring for 2 h at 5 – 10 ℃, and poured into a beaker contained 250 g of chipped ice. The mixture is diluted with ice – water to a total volume of 750 ml and subjected to a filtration. The filter cake is then washed with waterand suspended in 40 ml of 2 % aqueous solution of sodium carbonate. After thorough mixing, the crude product *p* – nitrophenylmethylene diacetate is collected by filtration and washed with water and ethanol.

A 250 ml, round – bottomed flask is charged with the crude product obtained above, 20 ml of water, 20 ml of alcohol and 2 ml of concentrated sulfuric acid. The resultingmixuture is refluxed for 30 min, tand filtered as soon as possible. The filtrate is chilled in a cool water bath to give a crystal, which is collected by filtration and washed with water washing . m. p 106 ~ 107 ℃.

【Note】

1. CAUTION: It is important to add chromium trioxide into the acetic anhydride gradually, a reverse charging sequence may cause an explosion.

2. More crystal of product will form if the filtrate is further diluted with 50 ml of water.

【Subjects for Thinking】

1. What is the reason that this oxidation process must go through the intermediate step of *p* – nitrobenzaldiacetate diacetate?

2. In addition tochromium trioxide and acetic anhydride, what other oxidation reagents can also be used in this preparation?

实验十七 二苯甲烷的制备

【实验目的】

1. 学习转移氢化反应的机理、氢源的种类及特点。

2. 了解氢化反应催化剂的种类及应用。

【反应式】

O

Ni,*i*-PrOH

【实验试剂】

二苯酮：2. 5 g

Raney 镍：2. 5 g

异丙醇：50 ml

【操作步骤】

在配有搅拌子、温度计和回流冷凝器的250 ml 三颈瓶中，加入二苯酮2. 5 g 和含有2. 5 g Raney 镍的50 ml 异丙醇悬浮液，搅拌下回流反应1 h。冷却至室温，倾倒出有机溶剂，残留的Raney 镍用异丙醇淋洗3 次（10 ml ×3）。合并后的有机溶液用硅藻土过滤，滤液减压浓缩除去溶剂，得到无色油状物即为二苯甲烷粗品，得量2. 2 g，收率96. 0 %，沸点265 ℃。

【注意事项】

1. Raney 镍会吸附到搅拌子上，这有利于其分离。因Raney 镍性质活泼，易燃，处理后应妥善保存。

2. 催化剂当在干燥条件下暴露于空气中时非常容易自燃，应当一直保存在溶剂中保持湿润。

【思考题】

1. 本反应的反应机理如何？

2. 异丙醇在该反应中的作用有哪些？是否可以采用其他试剂代替？

Experiment 17 The preparation of diphenylmethane

【Aim】

1. To learn the mechanisms of Transfer Hydrogenation, the categories and characteristics of hydrogenation sources.

2. To understand categories and applications of hydrogen catalysts.

【Reaction equation】

Benzophenone $\xrightarrow{\text{Ni}, i\text{-PrOH}}$ Diphenylmethane

【Reagents】

Benzophenone: 2. 5 g

Raney Ni: 2. 5 g

2 – Ppropanol: 50 ml

【Procedure】

A 250 ml, three – necked flask equipped with a stirring bar, a thermometer and a reflux condenser, is charged with 2. 5 g of benzophenone and a suspension solution of 2. 5 g of Raney nickel in 50 mL of 2 – propanol. After the mixture is refluxed for 1 h and cooled to room temperature, the catalyst is removed by filtration and washed with 2 – propanol（10 mL ×3）. The combined organic layers are filtered through celite and concentrated by rotary evaporation to yield 2. 2 g of colorless transparent oil as the crude product diphenylmethane（96. 0%）. b. p 265 ℃.

【Note】

1. Raney nickel is paramagnetic and clings to the magnetic stir bar, which facilitates decantation.

2. CAUNTION：The catalyst is extremely pyrophoric when exposed to the air in a dry condition；it should be kept wet with solvent at all times.

【Subjects for Thinking】

1. What is the mechanism of this reaction?

2. What are functions of 2 – propanol in this reaction? Whether it could be replaced with other regents or not?

实验十八　二苯甲醇的制备

【实验目的】

学习酮的还原反应机理、还原剂的种类及特点。

【反应式】

$$\text{PhCOPh} \xrightarrow[C_2H_5OH]{NaBH_4} \text{PhCH(OH)Ph}$$

【实验试剂】

二苯酮：5. 5 g

硼氢化钠：0. 6 g

乙醇（95 %）：30 ml

【操作步骤】

在配有搅拌子、温度计、回流冷凝器的 250 ml 的三颈瓶中，加入二苯酮 5. 5 g、95% 乙醇 30 ml，水浴加热至反应物全溶。冷却至室温，搅拌下分批加入硼氢化钠 0. 6 g，加入速度以使反应温度保持在 50 ℃以下为宜。加毕，回流反应 1 h。冷却到室温，加入 30 mL 水稀释反应液，再加入 10 % 稀盐酸分解未反应的硼氢化钠，待冷却到室温后抽滤，水洗滤饼，抽干得粗产品。以石油醚（沸点 30 ~ 60 ℃）为溶剂重结晶可制得精品。

【思考题】

1. 除了本实验提供的方法外可否采用其他反应途径制备二苯甲醇?

2. 反应中的硼氢化钠为什么要分批加入?

Experiment 18 The preparation of Diphenyl carbinol

【Aim】

To comprehend the mechanism of reduction of ketone, and the categories and characteristics of reduction reagents.

【Reaction equation】

Ph–CO–Ph $\xrightarrow[C_2H_5OH]{NaBH_4}$ Ph–CH(OH)–Ph

【Reagents】

Diphenyl ketone: 5. 5 g

Sodium borohydride: 0. 6 g

Ethanol (95 %): 30 ml

【Procedure】

A 250 ml three – necked flask equipped with a stirring bar, a thermometer and a reflux condenser, is charged with 5. 5 g of diphenyl ketone and 30 ml of ethanol (95 %). The mixture is heated in a water bath till all solid dissolve completely. After the temperature is adjusted back to room temperature, 0. 6 g of sodium borohydride is added in batches so as to maintain the intrenal temperature below 50 ℃. After complete addition, the mixture is refluxed for 1 h and cooled to room temperature, at which point 30 ml of cold water is added. Then, 10 % hydrochloric acid is added into the reaction mixture to decompose excessive sodium borohydride. The crude product is colletcted by filtration and washed with water. A recrystallization from petroleum ether (b. p 30 ~60 ℃) gives the fine product.

【Subjects for Thinking】

1. What other methods can be used for this preparation?

2. Why the sodium borohydride should be added in batches in this procedure?

实验十九 对硝基乙酰苯胺的制备

【实验目的】

学习硝化反应的机理、硝化剂的种类及其特点。

【反应式】

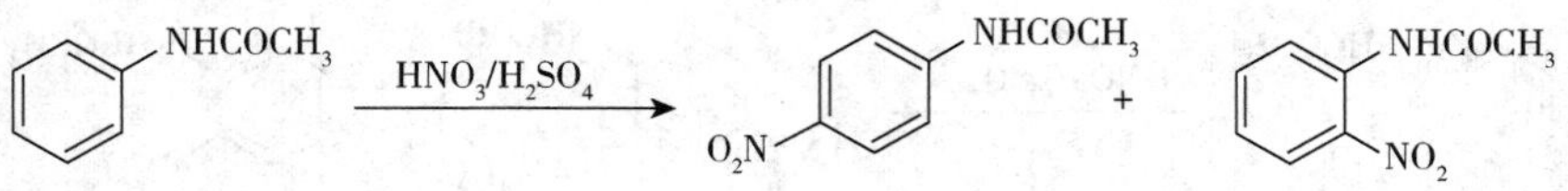

$$C_6H_5NHCOCH_3 \xrightarrow{HNO_3/H_2SO_4} p\text{-}O_2NC_6H_4NHCOCH_3 + o\text{-}O_2NC_6H_4NHCOCH_3$$

【实验试剂】

乙酰苯胺：13.5 g

冰醋酸：13.5 ml

浓硫酸：27 ml +6 ml

浓硝酸：6.9 ml

乙醇（95%）：130 ml

冰：130 g

【操作步骤】

在配有搅拌子、温度计、回流冷凝器及滴液漏斗的 250 ml 三颈瓶中，加入乙酰苯胺 13.5 g 及冰醋酸 13.5 ml。开动搅拌，在水浴冷却下滴加浓硫酸 27 ml，滴加过程中保持反应温度不超过 30 ℃。冰盐浴冷却此反应液至 0 ℃，滴加配制好的混酸（由浓硫酸 6 ml 和浓硝酸 6.9 ml 配制而成），滴加过程中严格控制滴加速度使反应温度不超过 10 ℃，滴加完毕，于室温下放置 1 h。将反应混合物在搅拌下倒入装有 130 g 碎冰的烧杯中，即刻有黄色沉淀析出，待碎冰全部融化后抽滤，冰水洗涤滤饼至洗水呈中性，抽干得粗品。将该粗品以 130 ml 乙醇重结晶，得对硝基乙酰苯胺精品 9 ~ 11 g，熔点 213 ~ 214 ℃。

【注意事项】

1. 加入硫酸时激烈放热，因此需慢慢加入，此时反应液应为澄清液。
2. 配制混酸时放热，要在冷却及搅拌条件下配制，要将硫酸逐滴加到硝酸中去。

【思考题】

1. 本实验中采用乙醇重结晶法分离邻、对位硝化产物的根据是什么？
2. 冰解一步的原理是什么？
3. 配制混酸过程中有时制得的混酸带有浅棕色，分析其原因。

Experiment 19 The preparation of *p* – nitroacetanilide

【Aim】

To comprehend the mechanism of nitration reaction and typical nitration reagents.

【Reaction equation】

$$\text{C}_6\text{H}_5\text{NHCOCH}_3 \xrightarrow{HNO_3/H_2SO_4} p\text{-}O_2N\text{C}_6\text{H}_4\text{NHCOCH}_3 + o\text{-}NO_2\text{C}_6\text{H}_4\text{NHCOCH}_3$$

【Reagents】

Acetanilide: 13. 5 g

Glacial acetic acid: 13. 5 ml

Concentrated sulfuric acid: 33 ml

Nitric acid (65% ~69%): 6. 9 ml

Ethanol (95%): 130 ml

【Procedure】

A 250 ml, three – necked flask immersed in an ice – salt bath and equipped with a stirring bar, a thermometer, a reflux condenser and a dropping funnel, is charged with 13. 5 g of acetanilide and 13. 5 ml of acetic acid. 27 ml of concentrated sulphuric acid is then added dropwise through the dropping funnel at such a rate that the temperature does not exceed 30 ℃. A cold mixture of 6. 9 ml of concentrated sulphuric acid and 6 ml of nitric acid is added dropwise at such a rate that the temperature does not exceed 10 ℃ . After complete addition, the mixture is allowed standing for 1 h at room temperature, then poured into a beaker containing 130 g of crushed ice. After the ice has all melted, the crude nitroacetanilide as the precipitate is collected by filtration and washed with water until the washings are free of acid. The crude material is pressed free of excess water and recrystallized from 130 ml of ethanol, giving 9 ~ 11 g of the fine product of *p* – nitroacetanilide as a colorless, crystalline solid. m. p 213 ~214 ℃.

【Note】

1. The process is an exothermic reaction, so the concentrated sulphuric acid should be added gradually.

2. The sulphuric acid must be added to the nitric acid when mixing the two acids under cooling and stirring.

【Subjects for Thinking】

1. What is the reason that *o* – nitroacetanilide and *p* – nitroacetanilide isomers can be separated by recrystallization from alcohol?

2. What is the principle for adding the mixture into the ice solution?

3. What is the reason that sometimes the mixed acid produced has a light brown color?

第二部分 综合性实验

实验二十 苯佐卡因的制备

【中文名称】苯佐卡因

【英文名称】Benzocaine

【化学名】对氨基苯甲酸乙酯

【结构式】

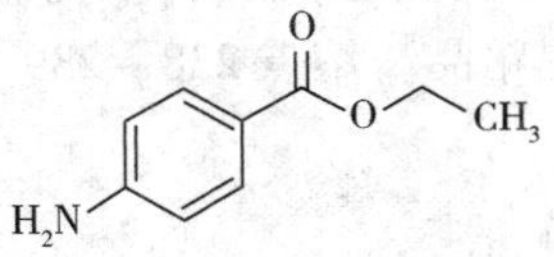

$$H_2N-C_6H_4-COOCH_2CH_3$$

【分子式】$C_9H_{11}NO_2$

【分子量】165.19

【理化性质】本品为白色结晶性粉末；无臭，味微苦，随后有麻痹感；遇光色渐变黄；易溶于醇、乙醚、三氯甲烷，在脂肪油中略溶，在水中极微溶解；在稀酸中溶解，熔点：89 ~ 92 ℃。

【临床应用】麻醉药与麻醉辅助用药 - 局麻药。

【合成路线】

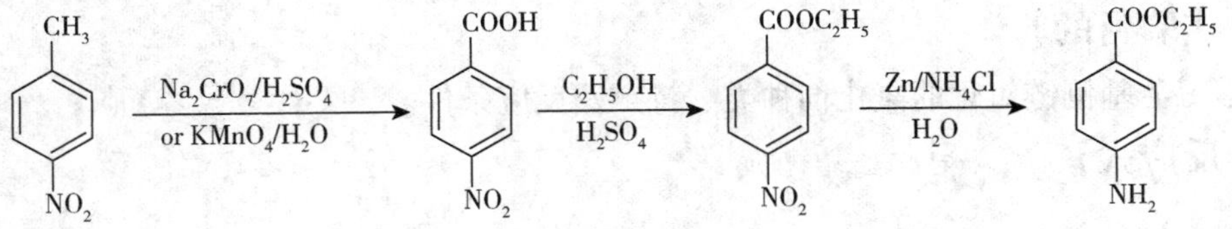

$$p\text{-}NO_2C_6H_4CH_3 \xrightarrow[\text{or } KMnO_4/H_2O]{Na_2CrO_7/H_2SO_4} p\text{-}NO_2C_6H_4COOH \xrightarrow[H_2SO_4]{C_2H_5OH} p\text{-}NO_2C_6H_4COOC_2H_5 \xrightarrow[H_2O]{Zn/NH_4Cl} p\text{-}NH_2C_6H_4COOC_2H_5$$

一、对硝基苯甲酸的制备（$Na_2Cr_2O_7$法）

【实验目的】

学习芳基侧链氧化的氧化剂种类、特点及反应条件。

【反应式】

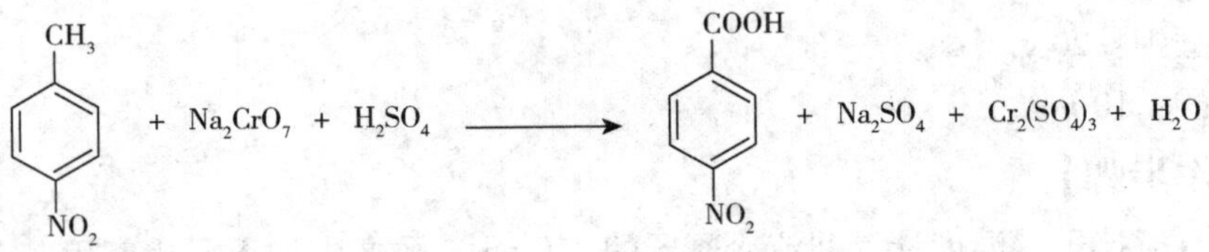

$$p\text{-}NO_2C_6H_4CH_3 + Na_2CrO_7 + H_2SO_4 \longrightarrow p\text{-}NO_2C_6H_4COOH + Na_2SO_4 + Cr_2(SO_4)_3 + H_2O$$

【实验试剂】

对硝基甲苯：6 g

重铬酸钠：18 g

浓硫酸：28 ml

【操作步骤】

在装有搅拌子、温度计、回流冷凝器和滴液漏斗的250 ml三颈瓶中，加入对硝基甲苯6 g、重铬酸钠18 g及水40 ml，开动搅拌，将28 ml浓硫酸由滴液漏斗加入到反应瓶中，加毕，在沸水浴上加热至80 ℃反应1.5 h，冷却到室温，加入50 ml水。抽滤，滤饼用50 ml水洗涤两次。将滤饼转移到100 ml的圆底烧瓶中，加入5 %的稀硫酸25 ml，于沸水浴上加热10 min，冷却到室温后抽滤，将滤饼溶于约30 ml 5 %的氢氧化钠中，再加入活性炭0.3 g，加热至50 ℃，脱色5 min后，趁热抽滤，将滤液在搅拌下慢慢倾入60 ml 15 %的硫酸中得浅黄色沉淀，抽滤，水洗滤饼，抽干得产品，收率82 %。以乙醇为溶剂重结晶可得精品，熔点238～239 ℃。

【注意事项】

1. 滴加硫酸时温度不能超过30℃。
2. 注意控制反应温度，温度过高时对硝基甲苯易升华而结晶于冷凝器底部。
3. 此步是为了溶解铬盐。

【思考题】

1. 本实验是否可选其他氧化剂?
2. 分析本实验后处理过程中产品的分离原理。

二、对硝基苯甲酸的制备（$KMnO_4$法）

【实验目的】

学习芳基侧链氧化的氧化剂种类、特点及反应条件。

【反应式】

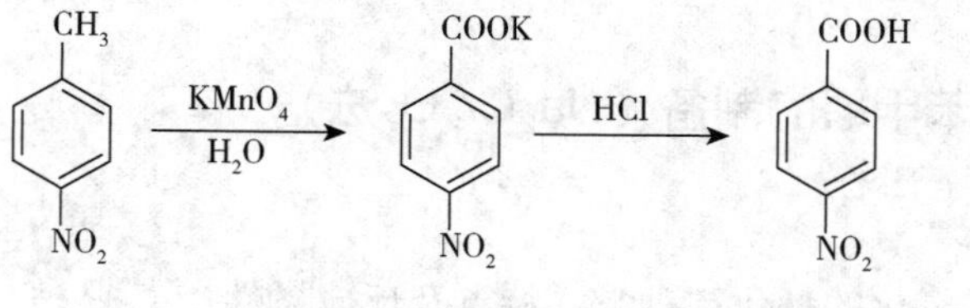

$$p\text{-}CH_3C_6H_4NO_2 \xrightarrow[H_2O]{KMnO_4} p\text{-}KOOCC_6H_4NO_2 \xrightarrow{HCl} p\text{-}HOOCC_6H_4NO_2$$

【实验试剂】

对硝基甲苯：7 g

高锰酸钾：20 g

浓盐酸：10 ml

【操作步骤】

在装有搅拌子、温度计、回流冷凝器的250 ml三颈瓶中加入对硝基甲苯7 g、高锰酸钾 10 g 及水 100 ml，开动搅拌在沸水浴上加热至 80 ℃反应 1 h，加入高锰酸钾 5 g，反应 1 h 后再加入高锰酸钾5 g，反应 0.5 h 后升温至水浴沸腾，继续反应直到高锰酸钾的颜色完全消失。冷却至室温，抽滤，20 ml 水洗涤滤饼一次，滤液在搅拌下加 10 ml 浓盐酸酸化，待析出的沉淀冷却至室温后抽滤，水洗滤饼，抽干得粗产品。以乙醇为溶剂重结晶可得精品，熔点 238 ~239 ℃。

【注意事项】

1. 注意控制反应温度，温度过高时对硝基甲苯易升华而结晶于冷凝器底部。
2. 高锰酸根带有鲜红色，而生成的二氧化锰为黑色沉淀。

【思考题】

1. 实验中高锰酸钾为什么要分批加入?
2. 试比较该方法与重铬酸钠氧化方法的优缺点?

三、对硝基苯甲酸乙酯的制备

【实验目的】

学习酯化反应的特点及反应条件。

【反应式】

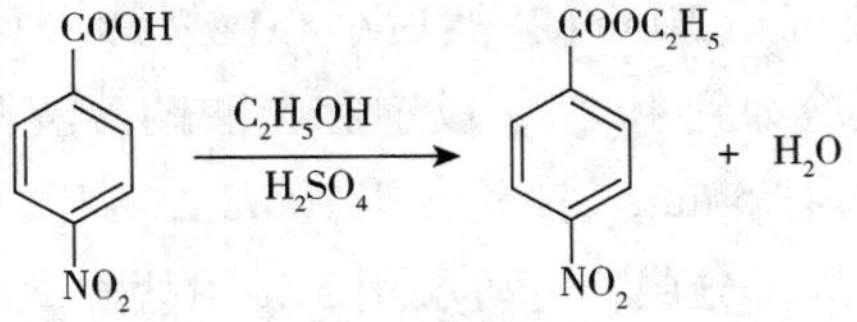

$$p\text{-}NO_2C_6H_4COOH \xrightarrow[H_2SO_4]{C_2H_5OH} p\text{-}NO_2C_6H_4COOC_2H_5 + H_2O$$

【实验试剂】

对硝基苯甲酸：12 g

无水乙醇：50 ml

浓硫酸：6 ml

【操作步骤】

于 100 mL 的圆底烧瓶中依次加入对硝基苯甲酸 12 g、无水乙醇 50 ml 及浓硫酸 6 ml，装上回流冷凝器在水浴上加热回流反应 2 h，冷却至室温后，常压蒸出部分乙醇（7 ~10 ml），将反应物倒入 120 ml 冷水中，析出白色沉淀，抽滤，将滤饼转移到 100 ml的烧杯中，加入 5 % 的碳酸钠溶液 20 ml 搅拌片刻，抽滤，水洗滤饼至洗水呈中性，抽干得产品，熔点 54 ~56 ℃。

【思考题】

1. 反应中加入浓硫酸的目的是什么?
2. 在操作中为什么要蒸出部分乙醇?
3. 后处理中加入 5 % 的碳酸钠溶液的作用是什么?

四、苯佐卡因的制备

【实验目的】

学习“锌－电解质”还原硝基的原理及基本反应条件。

【反应式】

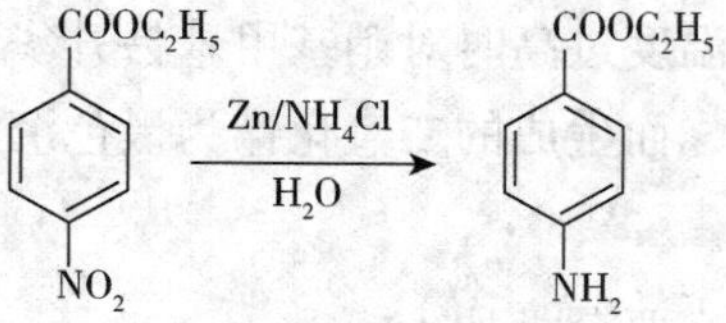

$$p\text{-}NO_2C_6H_4COOC_2H_5 \xrightarrow[H_2O]{Zn/NH_4Cl} p\text{-}NH_2C_6H_4COOC_2H_5$$

【实验试剂】

对硝基苯甲酸乙酯：6 g

氯化铵：2 g

锌粉：10 g

【操作步骤】

在 250 ml 的圆底烧瓶中依次加入 2.0 g 氯化铵、10 g 锌粉和 50 ml 水，开动搅拌，沸水浴加热至 95 ℃，活化锌粉 10 min，然后小心慢慢加入 6 g 对硝基苯甲酸乙酯，于 95 ℃反应 60 min，反应液冷却到 40 ℃，以饱和碳酸钠溶液调反应液 pH = 7 ~ 8，再加入 30 ml 二氯甲烷，充分搅拌 5 min，抽滤，用 10 ml 二氯甲烷洗涤滤饼。滤液和洗液转移至 250 ml 分液漏斗中，静置分层，有机层用 5 % 稀盐酸萃取（30 ml × 3），合并盐酸萃取液，并用 40 % 氢氧化钠溶液调至 pH = 8，析出结晶，抽滤，用少量水洗涤滤饼，干燥，得对氨基苯甲酸乙酯。

【注意事项】

1. 加入碳酸钠饱和溶液是为了游离反应生成的对氨基苯甲酸盐酸盐。
2. 在萃取过程中要注意及时排气。

【思考题】

1. 本实验中的硝基化合物是否可以采用其他的还原方法？
2. 锌－电解质还原反应机理是什么？
3. 绘出粗产品纯化过程及其原理的流程图。

Experiment 20 The preparation of Benzocaine

【Name】 Benzocaine

【Chemical Name】 Ethyl 4 – Aminobenzoate acid

【Structure】

O
O CH$_3$
H$_2$N

【Molecular Formula】 $C_9H_{11}NO_2$

【Molecular Weight】 165.19

【Properties】 White crystalline powder, barely odorless, slightly bitter taste, then numbness, more soluble in dilute acids and very soluble in ethanol, chloroform and ethyl ether, slightly soluble in fatty oil, sparingly soluble in water, m. p 89 ~ 92℃

【Application】 a local anesthetic

【Reaction equation】

CH_3 / NO_2 $\xrightarrow[\text{or } KMnO_4/H_2O]{Na_2CrO_7/H_2SO_4}$ COOH / NO_2 $\xrightarrow[H_2SO_4]{C_2H_5OH}$ $COOC_2H_5$ / NO_2 $\xrightarrow[H_2O]{Zn/NH_4Cl}$ $COOC_2H_5$ / NH_2

one The preparation of *p* – nitrobenzoic acid (with $Na_2Cr_2O_7$)

【Aim】

To comprehend the categories of oxidation reagents and their characteristics, reaction conditions for the oxidization of aromatic side chains.

【Reaction equation】

CH_3 / NO_2 + Na_2CrO_7 + H_2SO_4 $\longrightarrow$ COOH / NO_2 + Na_2SO_4 + $Cr_2(SO_4)_3$ + H_2O

【Reagents】

p – Nitrotoluene: 6 g

Sodium bichromate: 18 g

Concentrated sulfuric acid: 28 ml

【Procedure】

A 250 ml, three – necked flask equipped with a stirring bar, a thermometer, a dropping funnel and a reflux condenser, is charged with 6 g of *p* – nitrotoluene, 18 g of sodium bichromate and 40 ml of water. Then, 28 ml of concentrated sulfuric acid is added dropwise through the dropping funnel. After complete addition, the mixture is allowed stirring at 80 ℃ for 1.5 h.

After cooled to room temperature, the suspension is diluted with 50 mL of water, and the solid is collected by filtrationand washed with water (50 ml ×2) . The filter cake is transferred into a 100 ml, round – bottomed flask, and 25 ml of sulfuric acid (5 %) is added. The resulting mixture is heated in a boiling water bath for 10 min [3] and cooled back to room temperature. The insoluble solid is collected by filtration and dissolved in 30 mL of sodium hydroxide (5 %) . The resulting solution is treated with 0. 3 g of active carbon at 50 ℃ for 5 min. After filtration, the filtrate is poured into 60 ml of sulfuric acid (15 %) under stirring, and a quantity of sediment precipitates, which is collected by filtration and washed with water. The crude product is obtained after drying at room temperature to constant weight (82 % yield) . If required, the fine product may be recrystallized from ethanol as white powder. m. p 238 ~ 239 ℃.

【Note】

1. This procedure is exothermic reaction, so the addition of concentrated sulfuric acid should be controlled at such a rate that the internal temperature does notexcced 30 ℃.

2. *p* – Nitrotoluene will sublimate and crystallize on the bottom of the reflux condenser if the temperature is too high.

3. This opetation is to dissolve the chromium salts with diluted sulfuric acid.

【Subjects for Thinking】

1. What other oxidation reagents can be used in this preparation?

2. What is the separation principle of the target product?

two The preparation of *p* – nitrobenzoic acid (with $KMnO_4$)

【Aim】

To comprehend the categories of oxidation reagents and their characteristics, reaction conditions for the oxidization of aromatic side chains.

【Reaction equation】

$$p\text{-}CH_3C_6H_4NO_2 \xrightarrow[H_2O]{KMnO_4} p\text{-}KOOCC_6H_4NO_2 \xrightarrow{HCl} p\text{-}HOOCC_6H_4NO_2$$

【Reagents】

p – Nitrotoluene: 7 g

Potassium permanganate: 20 g

Concentrated hydrochloric acid: 10 ml

【Procedure】

A 250 ml, three – necked flask equipped with a stirring bar, a thermometer, and a reflux condenser, is charged with 7 g of *p* – nitrotoluene, 10 g of potassium permanganate and 100 mL of water. The resulting mixture is allowed stirring at 80 ℃ for 1 h, at which point 5 g of potassium permanganate is added, and the mixture is stirred for another 1 h. Then, one more portion of 5 g of potassium permanganate is added, the mixture is stirred for 0. 5 h and refluxed until the color of potassium permanganate disappears. After cooled to room temperature, the mixture is subjected to a filtration and the filter cake is washed with 20 mL of water. The filtrate and washings are acidified with 10 mL of concentrated hydrochloric acid to give a quantity of sediment. The crude product is collected by filtration and washed with water. If required, the fine product may be recrystallized from ethanol as white powder. m. p 238 ~ 239 ℃.

【Note】

1. *p* – Nitrotoluene will sublimate and crystallize on the bottom of the reflux condenser if the temperature is too high.

2. The color of permanganate is pink, while the manganese dioxide is a black precipitate.

【Subjects for Thinking】

1. What is the reason for adding the potassium carbonate in portions in this procedure?

2. What are the advantages and disadvantages of these two oxidation methods applying sodium bichromate and potassium permanganate?

three　The preparation of ethyl *p* – nitrobenzoate

【Aim】

To comprehend the characteristics, reaction conditions of esterification reactions.

【Reaction equation】

$$p\text{-}NO_2C_6H_4COOH \xrightarrow[H_2SO_4]{C_2H_5OH} p\text{-}NO_2C_6H_4COOC_2H_5 + H_2O$$

【Reagents】

p – Nitrobenzoic acid: 12 g

Anhydrous ethyl alcohol: 20 ml

Concentrated sulfuric acid: 6 ml

【Procedure】

A 100 ml, round – bottomed flask equipped with a stirring bar and a reflux condenser, is charged with 12 g of *p* – nitrobenzoic acid, 50 ml of anhydrous ethyl alcohol and 6 ml of concentrated sulfuric acid. The mixture is refluxed for 2 h and cooled to room temperature. Then part of alcohol is distilled off. The residue is poured into 120 ml of cold water, resulting in a white precipitate. The crude product is collected by filtration, washed with water, and transferred into 20 ml of 5 % sodium carbonate solution. After thorough mixing, the product is collected by filtration and washed with water until the washings are free of acid. The product is dried at room temperature to constant weight. m. p 54 ~ 56 ℃.

【Subjects for Thinking】

1. What is the reason for adding concentrated sulfuric acid to the reaction mixture?

2. What is the reason for the distillation of a small part of alcohol from the reaction mixture?

3. What is the reason for adding 5 % sodium carbonate (*a. q.*) into the crude product in the final treatment process?

four The preparation of Benzocaine

【Aim】

To comprehend the mechanisms of reduction reactions for nitro group using "zinc – electrolyte" system and understand the corresponding basic reaction conditions.

【Reaction equation】

$$p\text{-}O_2N\text{-}C_6H_4\text{-}COOC_2H_5 \xrightarrow[H_2O]{Zn/NH_4Cl} p\text{-}H_2N\text{-}C_6H_4\text{-}COOC_2H_5$$

【Reagents】

Ethyl 4 – nitrobenzoate: 6 g

Ammonium chloride: 2 g

Zinc powder: 10 g

【Procedure】

A 250 ml, round – bottomed flask fitted with a stirring bar and a reflux condenser is charged with 2 g of ammonium chloride, 10 g zinc powder and 50 ml of water. The resulting mixture is heated to 95 ℃ and stirred for 10 min, at which point 6 g of ethyl *p* – aminobenzoate is added slowly. The mixture is allowed stirring at 95 ℃ for 1 h, and then cooled to 40 ℃, charging with saturated Na_2CO_3 solution to adjust its pH value to 7 ~ 8. Then 30 ml of dichloromethane is added to above solution and stirred for 5 min. The mixture is filtrated and washed with 10 ml of dichloromethane. The filtrate and washings are transferred to a 250 ml separating funnel, and the organic layer was separated, extracted with 5 % hydrochloric acid (30 ml × 3), and charged with 40 % NaOH solution to adjust its pH value to 8. The precipitate is collected by filtration and washed with water. The product is dried at room temperature to constant weight.

【Notice】

1. The aim for adding Na_2CO_3 is to free the salt of *p* – aminobenzoic acid hydrochloride.

2. Pay attention to exhaust gas in time after the extraction operation.

【Subjects for Thinking】

1. Are there any other reduction methods can also be used to reduce nitro – group in this experiment?

2. What is the reaction mechanism of thezinc – electrolyte reduction?

3. Draw a flowchart to illustrate the purification procedure and principle of the refinement of the crude product.

实验二十一　L –（+）– 扁桃酸乙酯的制备

【中文名称】 L –（+）– 扁桃酸乙酯

【英文名称】 Ethyl mandelate

【化学名】 L –（+）– α – 羟基苯乙酸乙酯，L –（+）– Ethyl – α – hydroxyphenylacetate

【结构式】

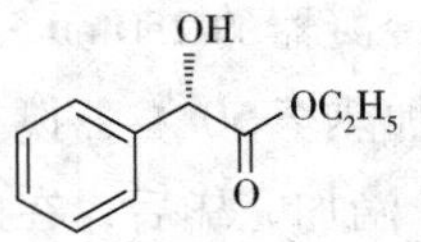

OH
OC$_2$H$_5$
O

【分子式】 $C_{10}H_{12}O_3$

【分子量】 180.2

【理化性质】 本品为无色透明液体，易溶于醇、乙醚、三氯甲烷等有机溶剂，在水中极微溶解；沸点 253 ~255 ℃。

【用途】 药物合成中间体，精细化学品。

【合成路线】

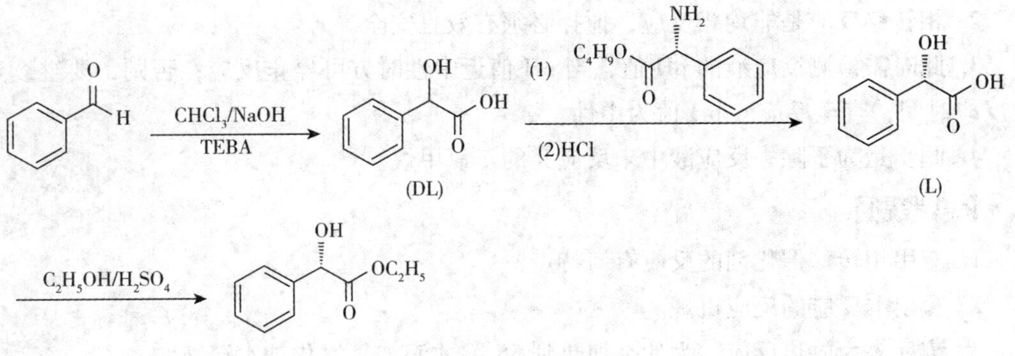

O
H
$CHCl_3/NaOH$
TEBA
OH
OH
O
(DL)
NH$_2$
(1)
C$_4$H$_9$O
O
(2)HCl
OH
OH
O
(L)
C_2H_5OH/H_2SO_4
OH
OC$_2$H$_5$
O

一、DL－扁桃酸制备

【实验目的】

1. 学习卡宾中间体进行的有机反应。

2. 学习相转移催化剂的原理和使用方法。

【反应式】

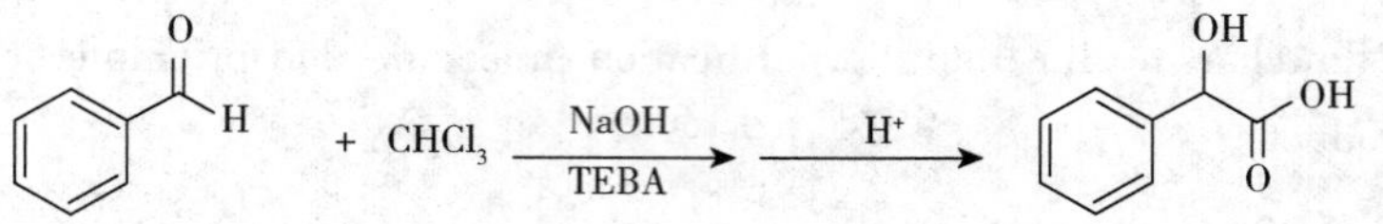

$$\text{C}_6\text{H}_5\text{CHO} + \text{CHCl}_3 \xrightarrow[\text{TEBA}]{\text{NaOH}} \xrightarrow{\text{H}^+} \text{C}_6\text{H}_5\text{CH(OH)COOH}$$

【实验试剂】

苯甲醛：6. 8 ml

三氯甲烷：12 ml

氢氧化钠/水：13 g/13 ml

苄基三乙基氯化铵（TEBA）：1 g

【操作步骤】

在装有搅拌子、温度计和回流冷凝器的250 ml三颈瓶中，依次加入6. 8 ml苯甲醛、1 g TEBA和12 ml三氯甲烷，水浴加热至50 ℃，搅拌下滴加氢氧化钠溶液，控制滴加速度使反应温度维持在55～65 ℃，滴加完毕后，在此温度下继续搅拌反应1 h。将反应液用140 ml水稀释，乙醚萃取反应液两次（15 ml×2），合并乙醚萃取液，倒入指定容器回收。水层用50％硫酸酸化至pH＝1，再用乙醚萃取两次（15 ml×2），合并酸化后的乙醚萃取液并以无水硫酸钠干燥过夜。

【注意事项】

1. TEBA可采取如下制备方法：在100 ml锥形瓶中加入5. 5 ml氯苄，7 ml三乙胺，19 ml 1，2－二氯乙烷，振摇均匀，室温放置一周，抽滤，干燥，得TEBA待用。

2. 相转移反应是非均相反应，搅拌必须有效且安全。

3. 此时需检测反应液的pH值，当pH值近中性时方可停止反应；否则，要继续延长反应时间，直至反应液pH值为中性。

4. 此步是为了除去反应液中未反应完的三氯甲烷。

【思考题】

1. 应用相转移催化剂的反应有何特点?

2. 写出本反应的反应机理。

3. 该反应为放热反应，为什么加热到55 ℃才滴加氢氧化钠水溶液?

4. 本实验中酸化前后分别用乙醚萃取，其目的是什么?

二、扁桃酸的精制

【实验目的】

学习混合溶媒为重结晶溶剂的原理及方法。

【实验试剂】

甲苯/无水乙醇混合溶剂（8∶1）：3 ml/1 g 粗品

【操作步骤】

以干燥的 100 ml 茄形瓶为蒸馏瓶，先常压蒸馏回收乙醚，再用水循环真空泵减压蒸尽残留的乙醚，蒸馏的残余物即为粗产品，得量6～7 g。将粗产物用甲苯－无水乙醇（8∶1）进行重结晶（每克粗产物约需 3 ml），趁热过滤，滤液在室温下放置则缓慢析出结晶，充分冷却后抽滤，即得精品扁桃酸，为白色结晶，熔点 118～119℃，干燥、称重，计算精制率和总产率。

【注意事项】

此步热过滤采用折叠的扇形滤纸法，玻璃漏斗需事先预热，以防止结晶析出。

【思考题】

重结晶过程选用混合溶剂的原因是什么？

三、扁桃酸的拆分（L－（＋）－扁桃酸的制备）

【实验目的】

学习拆分剂的使用原理及方法。

【反应式】

(DL) 扁桃酸 —— (1) D-苯甘氨酸丁酯（C_4H_9O-CO-CH(NH_2)-Ph）；(2) HCl ——→ (L) 扁桃酸

【实验试剂】

DL－扁桃酸：4.0 g

D－（－）－苯甘氨酸丁酯：5.3 g

盐酸（18 %）：100 ml

乙醚：360 ml

【操作步骤】

在配置有搅拌子和滴液漏斗的 100 ml 反应瓶中，依次加入 4.0 g DL－扁桃酸和

60 ml水，搅拌至全溶，室温搅拌下滴加5.3 g D－（－）－苯甘氨酸丁酯，滴加完毕，继续搅拌反应10 min，抽滤、滤饼干燥，得白色固体为D－苯丙氨酸丁酯L－扁桃酸盐，将其溶于100 ml 18 %稀盐酸中，室温下搅拌10 min，用乙醚萃取溶液三次（60 ml×3），合并乙醚萃取液，以无水硫酸镁干燥30 min，蒸除溶剂，得白色固体，干燥，得L－（+）－扁桃酸粗品。以水为溶剂重结晶，得L－（+）－扁桃酸精品，熔点132～134 ℃，$[\alpha]_D^{25}$ = +154（1 mol/L HCl）。

【思考题】

1. 本方法拆分DL－扁桃酸的原理是什么？
2. 除了使用拆分剂以外，还有其他什么方法用于制备光学纯的异构体？

四、L－（+）－扁桃酸乙酯的制备

【实验目的】

1. 学习酯化反应的特点及反应条件。
2. 学习共沸带水的基本原理及其在合成反应中的应用。

【反应式】

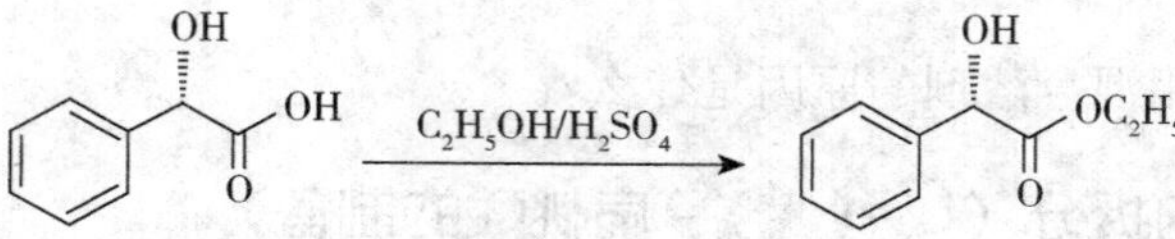

OH OH OH OC$_2$H$_5$ O O

C_2H_5OH/H_2SO_4

【实验试剂】

L－（+）－扁桃酸：10 g

无水乙醇：40 ml

浓硫酸：2 ml

【操作步骤】

在配置有搅拌子和回流冷凝器的100 ml圆底瓶中，加入L－（+）－扁桃酸10 g和无水乙醇40 ml，搅拌溶解后，滴入浓硫酸2 ml，回流反应2 h。减压浓缩出大部分溶剂，冷却，将残余物倾入50 g碎冰中，用饱和碳酸钠水溶液调pH＝8，二氯甲烷萃取三次（25 ml×3），用饱和氯化钠水溶液洗涤有机层，无水硫酸镁干燥。滤除干燥剂，浓缩回收二氯甲烷，向残留物中加入30 ml甲苯进行共沸带水以除去产品中残存的微量水，最终得产品为淡黄油状物11.2 g，收率94.6 %。

【思考题】

1. 该反应中加入少量浓硫酸的目的是什么？
2. 甲苯共沸带水的原理是什么？

Experiment 21 The preparation of L – (+) – ethyl mandelate

【Name】 Ethyl mandelate

【Chemical Name】 Ethyl 4 – Aminobenzoate acid

【Structure】

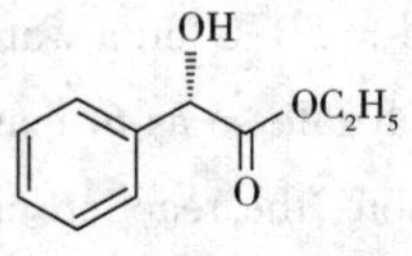

【Molecular Formula】 $C_{10}H_{12}O_3$

【Molecular Weight】 180. 2

【Properties】 Colorless liquid, soluble in alkahol, ether, and chloroform, slightly soluble in water. b. p 253 ~ 255. 2 ℃.

【Application】 Fine chemical for drug synthesis

【Reaction equation】

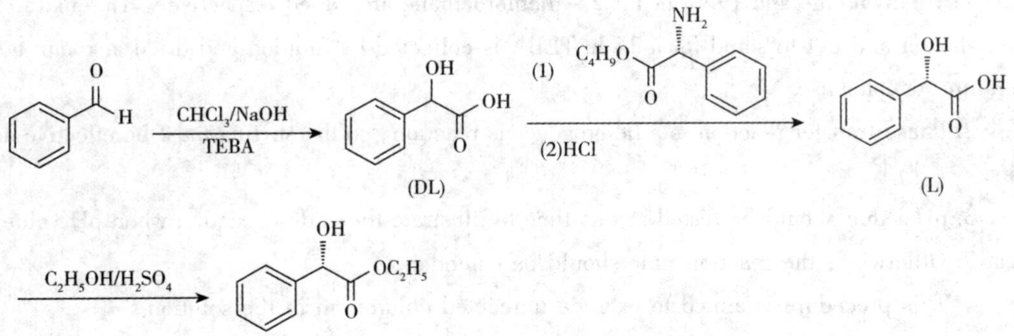

$CHCl_3/NaOH$, TEBA; (DL); (1) C_4H_9O (NH₂ ester), (2)HCl; (L); C_2H_5OH/H_2SO_4; OC_2H_5

one The preparation of DL – Mandelic Acid

【Aim】

1. To comprehend the characteristics of organic reactions involving carbine intermediates.
2. To comprehend the mechanism and the usage method of phase transfer catalyst.

【Reaction equation】

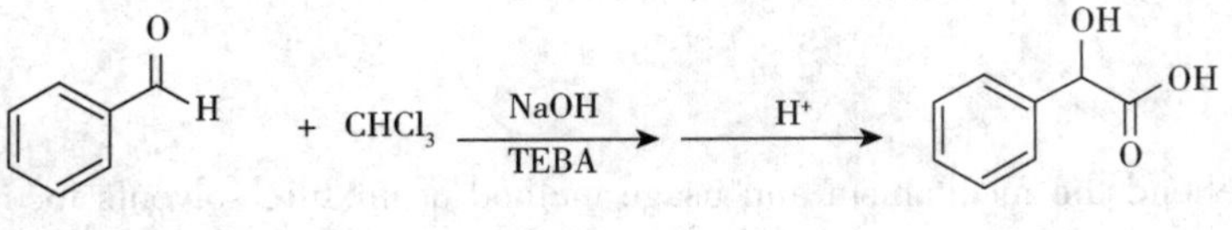

\+ $CHCl_3$ $\xrightarrow[TEBA]{NaOH}$ $\xrightarrow{H^+}$

【Reagents】

Benzaldehyde: 6. 8 ml

Chloroform：12 ml

Sodium hydroxide/Water：13 g/13 ml

Benzyltriethylaminium chloride（TEBA）：1 g

【Procedure】

A 250 – ml, three – necked flask equipped with a magnetic stirring bar, a thermometer and a reflux condenser, is charged with 6.8 ml of benzaldehyde, 1 g of TEBA and 12 ml of chloroform. The solution is then heated to 50 ℃ on a water bath, at which point sodium hydroxide solution is added dropwise at a such rate so as to maintain the internal temperature between 55 ℃ and 65 ℃. After complete addition, the reaction mixture is allowed stirring for 1 h at this temperature, then diluted with 140 mL of water and extracted with diethyl ether（15 ml ×2）. The organic layers are combined and transferred to designated containers for recovery. The water layer is adjusted to pH = 1 with 50 % sulfuric acid then extracted with diethyl ether（15 ml ×2）. The organic layers are combined and dried over anhydrous sodium sulfate overnight.

【Notice】

1. Synthetic procesure of TEBA：To a 100 ml conical flask, 5.5 ml of benzyl chloride, 7 ml of triethylamine and 19 ml of 1，2 – dichloroethane are added respectively. The mixture is well shaken and put to stand for a 24 h. TEBA is collected by filtation and dried at romm temperature for later use.

2. Phase transfer reaction is a heterogeneous reaction, so the stirring must be effective and safe.

3. pH value should be tested at this time to illustrate the end of reaction when pH value is near 7. Otherwise, the reaction time should be extended.

4. This procedure is aimed to exclude unreacted chloroform in the solution.

【Subjects for Thinking】

1. What are the characteristics of the reactions involving phase transfer catalysts?

2. What is the reaction mechanism?

3. This reaction is an exothermic reaction. What is the reason for not adding sodium hydroxide to the solution until the temperature reaches to 55 ℃?

4. What are the aims for extractions with diethyl ether before and after acidification?

two The recrystallization of Mandelic acid

【Aim】

To comprehend the mechanism and usage method of mixture solvents for recrystallization.

【Reagents】

2 Toluene/ethyl alcohol（8∶1）：3 ml /1 g crude product

【Procedure】

A 100 ml, pre – dried, round – bottomed flask is used as the distillation flask, diethyl ether is firstlycollected by an atmospheric distillation, and the residual diethyl ether is distilled completely by a reduced – pressure distillation to give 6 ~ 7 g of crude product. The crude product is then recrystallized from toluene – ethyl alcohol (8 : 1) (3 mL mixed solvent for 1 g crude). The resulting misture is refluxed and subjected to a hot filtration. The filtrate is chilled to room temperature, and the product precipitates, which is collected by filtration and dried at room temperature to constant weight. m. p 118 ~ 119 ℃.

【Notice】

This procedure applies a fan – shaped filter paper for hot filtration. Thus, the glass – fritted funnel should be preheated. If room – temperature glassware is used for this procedure, the product precipitates.

【Subjects for Thinking】

Why is a mixed solution of toluene/ethanlo chosen for recrystallization?

three　The resolution ofmandelic acid (The preparation of L – (+) – mandelic acid)

【Aim】

To comprehend the principle and application of resolution agents.

【Reaction equation】

(DL) mandelic acid —(1) C_4H_9O–C(=O)–CH(NH_2)–Ph; (2) HCl→ (L) mandelic acid

【Reagents】

DL – mandelic acid: 4. 0 g

D – (–) – Butyl phenylglycinate: 5. 3 g

18 % hydrochloric acid: 100 ml

diethyl ether: 360 ml

【Procedure】

A 100 ml, round – bottomed flask fitted with a stirring bar and a dropping funnel is charged with 4. 0 g of DL – mandelic acid and 60 ml of water. The solution is and stirred at room temperature till the solids fully dissolve. 5. 3 g of D – (–) – butyl phenylglycinate is added dropwise under stirring. After complete addition, the mixture is allowed stirring under room temperature for 10 min. A white solid, namely D –（–） – butyl phenylglycinate L – mandelate, is obtained by filtration, which is dissovled to 100 mL of 18 % hydrochloride acid and stirred for 10 min. The solution is extracted with dimethyl ether（60 ml ×3）, and the organic layers are combined and dried over anhydrous magnesium sulfate for 30 min. After filtration, the filtrate is evaporated off to give the crude product. A recrystallization from water gives the fine product. m. p 132 ~134 ℃, $[\alpha]_D^{25}$ = + 154（1 mol/L HCl）.

【Subjects for Thinking】

1. What is the principle of there solution of mandelic acid in this experiment?

2. What are the other methods that can be used to obtain optical pure isomers in addition to using resolving agents?

four The preparation of L –（+）– ethyl mandelate

【Aim】

1. To comprehend the characteristics, reaction conditions of esterification reaction.

2. To learn the principle of azeotropic distillation process and its synthesis applications.

【Reaction equation】

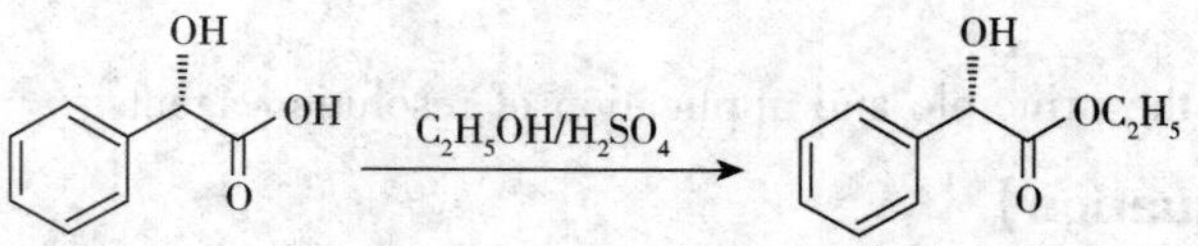

【Reagents】

L –（+）– Mandelic acid: 10 g

Anhydrous ethyl alcohol: 40 ml

Concentrated sulfuric acid: 2 ml

【Procedure】

A 100 ml, round – bottomed flask equipped with a stirring bar and a reflux condenser is charged with 10 g of L –（+）– mandelic acid, 40 ml of anhydrous ethyl alcohol and 2 ml of concentrated sulfuric acid. The mixture is refluxed on a boiling – water bath for 2 h, then concentrated by distillation. The residue is poured into 50 g of crushed ice, and adjusted to pH = 8 with saturated aqueous solution of sodium carbonate. The mixture is extracted with dichloromethane（25 ml ×3）. The combined organic layers are washed with saturated aqueous solution of sodium chloride（25 ml）, and dried over anhydrous magnesium sulfate. After filtration, the filtrate is evaporated off to give a residue, which is charged with 30 mL of toluene for an azeotropic distillation to give the target product as a yellow oil（11. 2 g, 94. 6 % yield）.

【Subjects for Thinking】

1. What is the reason for adding concentrated sulfuric acid to the reaction mixture?

2. What is the principle of azeotropic distillation?

实验二十二　布洛芬的制备

【中文名称】布洛芬

【英文名称】Ibuprofen

【化学名】2 -（4 - 异丁基苯基）丙酸，2 -（4 - isobutylphenyl）propionic acid

【结构式】

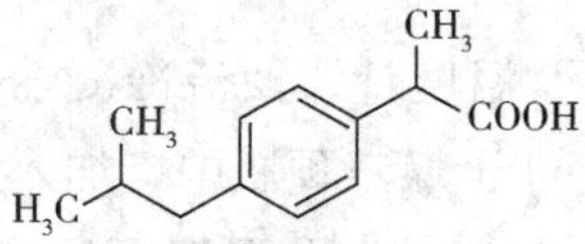

【分子式】$C_{13}H_{18}O_2$

【分子量】206. 28

【理化性质】本品为白色结晶性粉末，稍有特异臭；几乎不溶于水，可溶于丙酮、乙醚、二氯甲烷，可溶于氢氧化钠或碳酸钠水溶液；熔点：75 ~ 78 ℃。

【临床应用】非甾体抗炎药，广泛用于类风湿关节炎、风湿性关节炎等

【合成路线】

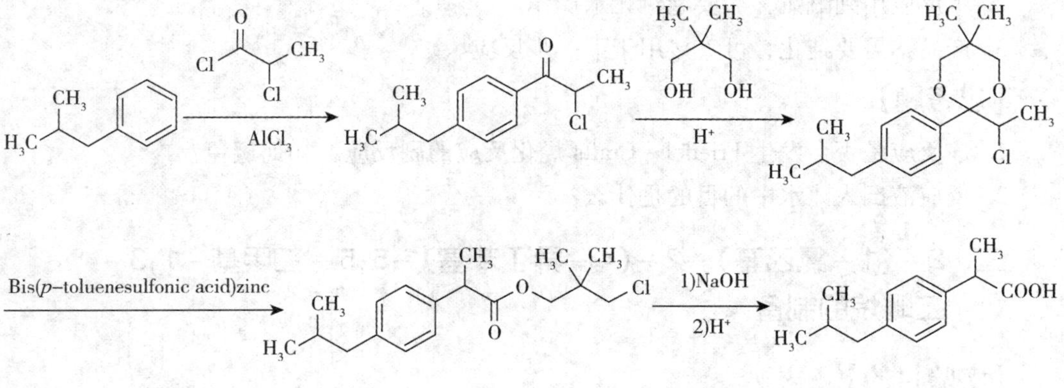

一、2 - 氯 - 1 -（4 - 异丁苯基）丙 - 1 - 酮的制备

【实验目的】

学习 Friedel - Crafts 酰化反应的机理和反应条件。

【反应式】

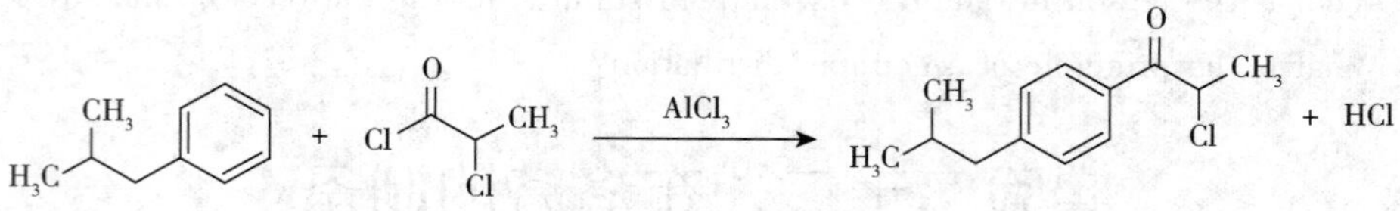

$$\mathrm{H_3C\!-\!CH(CH_3)\!-\!CH_2\!-\!C_6H_5} + \mathrm{Cl\!-\!C(=O)\!-\!CH(Cl)\!-\!CH_3} \xrightarrow{AlCl_3} \mathrm{H_3C\!-\!CH(CH_3)\!-\!CH_2\!-\!C_6H_4\!-\!C(=O)\!-\!CH(Cl)\!-\!CH_3} + HCl$$

【实验试剂】

无水三氯化铝：7. 5 g

2 - 氯丙酰氯：6. 3 g

异丁基苯：6. 4 g

石油醚（沸点 30 ~ 60 ℃）：30 ml

【操作步骤】

在配置有搅拌子、滴液漏斗、温度计、回流冷凝器（顶端装有无水氯化钙干燥管及氯化氢气体吸收装置）的 250 ml 干燥三颈瓶中，加入无水三氯化铝 7. 5 g，控温 20 ~ 35 ℃下滴加 2 - 氯丙酰氯 6. 3 g，加毕，在此温度下反应 20 min。向反应瓶中滴加异丁基苯 6. 4 g，控制滴加速度以保证反应温度在 15 ~ 30 ℃，滴毕，在此温度下继续反应 1 h。向反应物中加入石油醚 30 ml，搅拌 30 min，将反应液倒入 60 g 冰水中，于 25 ~ 35 ℃，搅拌 30 min。分出有机层，水层用石油醚提取三次（25 ml × 3），合并有机相，以水洗三次（30 ml × 3），即得到 2 - 氯 - 1 - （4 - 异丁苯基）丙酮（氯酮）的石油醚溶液。气相色谱法测含量，直接用于下一步反应。

【注意事项】

1. 称取三氯化铝时动作要快，要在通风橱内进行，以防其吸潮分解。
2. 本反应用到酰氯，需要绝对无水操作。
3. 产品不需要纯化，可直接用于下一步反应。

【思考题】

1. 该反应是否会发生 Friedel - Crafts 烃化反应的副反应？如何避免？
2. 反应液倒入冰水中的目的是什么？

二、2 - （1 - 氯乙基） - 2 - （4 - 异丁苯基） - 5，5 - 二甲基 - 1，3 - 二噁烷的制备

【实验目的】

学习制备缩酮的反应机理、特点和反应条件。

【反应式】

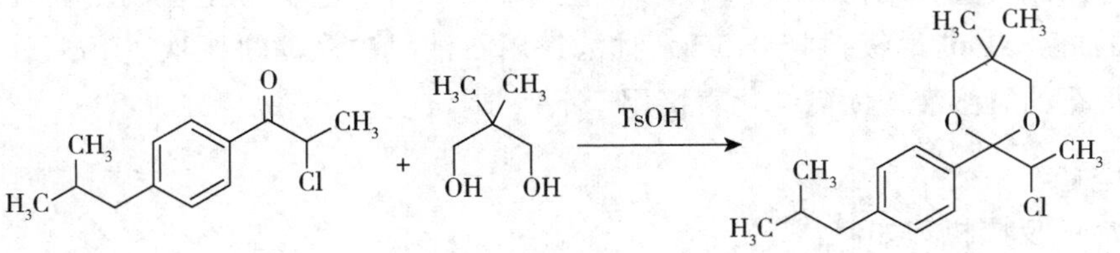

【实验试剂】

2-氯-1-(4-异丁苯基)丙-1-酮：前步制备所得的石油醚溶液

2，2-二甲基-1，3-二醇：6.9 g

对甲苯磺酸：0.5 g

【操作步骤】

在配置搅拌子、温度计和回流冷凝管的250 ml三颈瓶中，加入全部前步产物的石油醚溶液，2，2-二甲基-1，3-二醇6.9 g和水2 ml，搅拌下加入对甲苯磺酸0.5 g，升温至回流反应4 h。反应完毕，冷至0~5 ℃，过滤后即得到淡黄色的母液，为约含13 g 2-(1-氯乙基)-2-(4-异丁苯基)-5，5-二甲基-1，3-二噁烷(缩酮)的石油醚溶液，收率约95.0 %。

【思考题】

反应中加入对甲苯磺酸的作用是什么？

三、2-(4-异丁基苯基)丙酸-(3-氯-2,2-二甲基)丙酯的制备

【实验目的】

学习1，2-芳基迁移重排反应的机制、反应条件和特点。

【反应式】

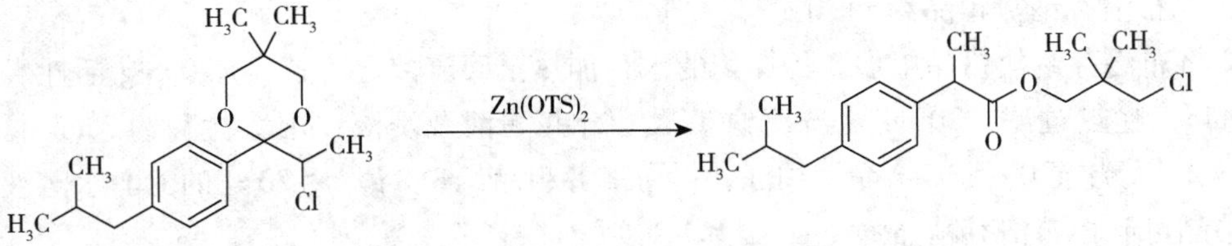

【实验试剂】

2-(1-氯乙基)-2-(4-异丁苯基)-5，5-二甲基-1，3-二噁烷：前步制备所得的缩酮-石油醚溶液。

对甲苯磺酸锌：0.6 g

硅藻土：0.5 g

【操作步骤】

在250 ml的三颈瓶上配置搅拌子、温度计及常压蒸馏装置。将上述缩酮石油醚液

全量，对甲苯磺酸锌 0.6 g 加入反应瓶中，搅拌下常压蒸馏去除石油醚溶剂。蒸毕，将蒸馏装置改为回流装置，维持 140～150 ℃反应 2.5 h。将反应液降至室温，加入硅藻土 0.5 g 和石油醚 10 ml 继续搅拌 0.5 h，抽滤，用石油醚洗涤滤饼两次（10 ml×2），合并滤液，得石油醚溶液，无需纯化直接用于下一步反应。

【思考题】

1. 本反应的反应机理如何？
2. 对甲苯磺酸锌在反应中有何作用？

四、布洛芬的制备

【实验目的】

学习酯水解反应的基本操作、反应条件及特点。

【反应式】

$$\text{CC(C)Cc1ccc(cc1)C(C)C(=O)OCC(C)(C)CCl} \xrightarrow[2)\,H^+]{1)\,NaOH} \text{CC(C)Cc1ccc(cc1)C(C)COOH} + \text{HOCC(C)(C)CCl}$$

【实验试剂】

氢氧化钠溶液 50 %：3.7 ml

浓盐酸：3 ml

【操作步骤】

在配置搅拌子、温度计、回流冷凝器的 250 ml 三颈瓶中加入上述重排产物的石油醚溶液，室温下缓慢滴加 50 % 的氢氧化钠液 3.7 ml，加毕，回流反应 40 min，向反应液中加入水 2 ml，冷却至 0～5 ℃放置析晶。抽滤，滤饼用石油醚洗涤两次（10 ml×2），得布洛芬钠盐粗品约 11.0 g。

将布洛芬钠盐粗品中加 8 倍体积的水，加热至回流使之溶解，加 0.5 g 活性炭脱色 30 min，趁热过滤，20 ml 水洗滤饼 1 次。合并滤液及洗液，用浓盐酸约 3 ml 酸化至 pH＝2，冷却至 0～5 ℃结晶。抽滤，得布洛芬粗品约 8.8 g。用 70 % 的乙醇水溶液进行重结晶可制得布洛芬精品。

【思考题】

本反应的实验操作机制如何？

Experiment 22　The preparation of Ibuprofen

【Name】 Ibuprofen

【Chemical Name】 2－（4－isobutylphenyl）propionic acid

【Structure】

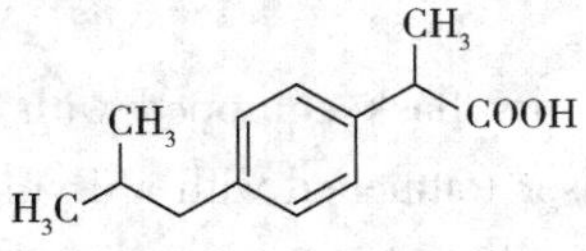

【Molecular Formula】$C_{13}H_{18}O_2$

【Molecular Weight】206. 28

【Properties】White crystalline powder, slight odour. Almost insoluble in water, soluble in acetone, ether, dichloromethane and sodium hydroxide or sodium carbonate aqueous solution, m. p 75 ~ 78 ℃

【Application】Nonsteroidal anti – inflammatory drug (NSAIDs), widely used in rheumatoid arthritis

【Reaction equation】

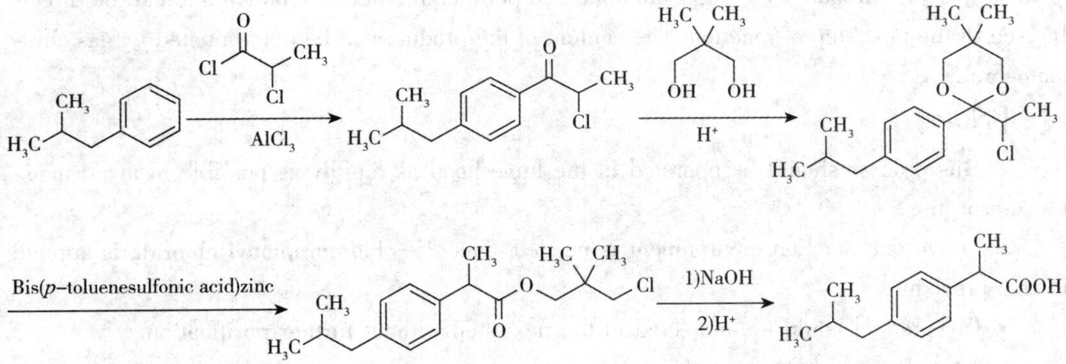

AlCl$_3$

H$^+$

Bis(*p*–toluenesulfonic acid)zinc

1)NaOH 2)H$^+$

one The preparation of 2 – chloro – 1 – (4 – isobutylphenyl) propan – 1 – one

【Aim】

To comprehend the mechanism, reaction conditions of Friedel – Crafts reaction.

【Reaction equation】

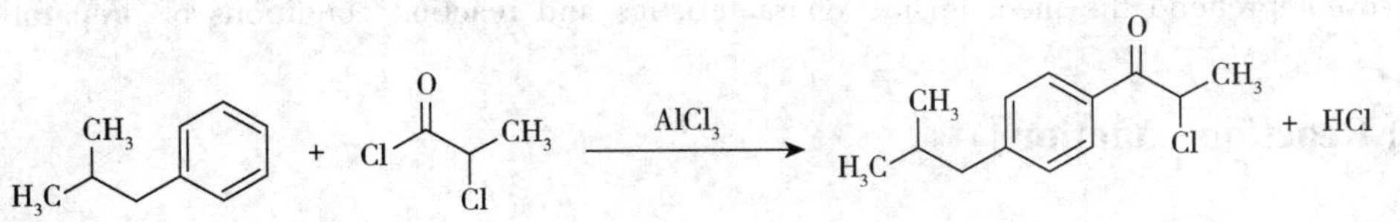

AlCl$_3$ + HCl

【Reagents】

Aluminium chloride: 7. 5 g

2 – Chloropropionyl chloride: 6. 3 g

Isobutylbenzene: 6. 4 g

Petroleum ether (b. p 30 ~ 60 ℃): 30 ml

【Procedure】

A 250 – ml, three – necked, dry flask equipped with a stirring bar, a thermometer, a dropping funnl and a reflux condenser (attached with a drying tube and a gas absorbent unit), is charged with 7. 5 g of aluminium chloride. Then, 6. 3 g of 2 – chloropropionyl chloride is added dropwise at such a rate so as to maintain the internal temperature between 20 ℃ and 35 ℃. After complete addition, the mixture is stirred at this temperature for 20 min, at which point 6. 4 g of isobutylbenzene is added dropwise to the mixture at such a rate that the temperature does not exceed 30 ℃. The resulting mixture is allowed strring for 1 h. Then, 30 ml of petroleum ether is added and the stirring is continued for 30 minutes. The mixture is then poured into 60 g of crushed ice – water, and stirred at 25 ~ 35 ℃ for 30 minutes. The organic layer is separated and the aqueous layer is extracted with petroleum ether (25 ml × 3). The combined organic layers are washed with water (30 ml × 3). The solution of 2 – chloro – 1 – (4 – isobutylphenyl) propan – 1 – one (chlorone) in petroleum ether is obtained and can be directly used in the next step of reaction. The content of the product can be determinated by gas chromatography.

【Note】

1. This process should be operated in the fume hood as rapidly as possible to avoid moisture absorption.

2. Anhydrous reaction environment is needed since 2 – chloropropionyl chloride is applied in this procedure.

3. The product could be subjected to the next step without further purification.

【Subjects for Thinking】

1. Does the side reaction of Friedel – Crafts alkylation happen? How to avoid it?

2. What is the purpose of pouring the reaction mixture into ice – water?

two The preparation of 2 – (1 – chloroethyl) – 2 – (4 – isobutylphenyl) – 5, 5 – dimethyl – 1,3 – dioxane

【Aim】

To comprehend the mechanism, characteristics and reaction conditions of preparation of ketal.

【Reaction equation】

H$_3$C CH$_3$ O CH$_3$ Cl + H$_3$C CH$_3$ OH OH —TsOH→ H$_3$C CH$_3$ O O CH$_3$ Cl CH$_3$ H$_3$C + H_2O

【Reagents】

1 – Chloroethyl – 4 – isobutylphenyl ketone: petroleum ether solution obtained in last step

2, 2 – Dimethylpropane – 1, 3 – diol: 6.9 g

p – Toluenesulfonic acid: 0.5 g

【Procedure】

A 250 ml, three – necked flask equipped with a stirring bar, a thermometer and a reflux condenser, is charged with the prepared solution of 2 – chloro – 1 – (4 – isobutylphenyl) propan – 1 – one (chlorone) in petroleum ether, 6.9 g of 2, 2 – dimethylpropane – 1, 3 – diol, 2 mL of water and 0.5 g of *p* – toluenesulfonic. The mixture is refluxed for 4 h, then cooled to 0 ~ 5℃. The resulting precipitation is filtered off and the solution of the crude product, 2 – (1 – chloroethyl) – 2 – (4 – isobutylphenyl) – 5, 5 – dimethyl – 1, 3 – dioxane (ketal), in petroleum ether (about 13 g, 95.0 % yield) is obtained as a yellow solution.

【Subjects for Thinking】

What is the reason for adding *p* – toluenesulfonic acid into the reaction?

three The preparation of 3 – chloro – 2,2 – dimethylpropyl 2 – (4 – isobutylphenyl) propanoate

【Aim】

To comprehend the mechanism, reaction conditions and characteristics of 1, 2 – aryl shift rearrangement reaction.

【Reaction equation】

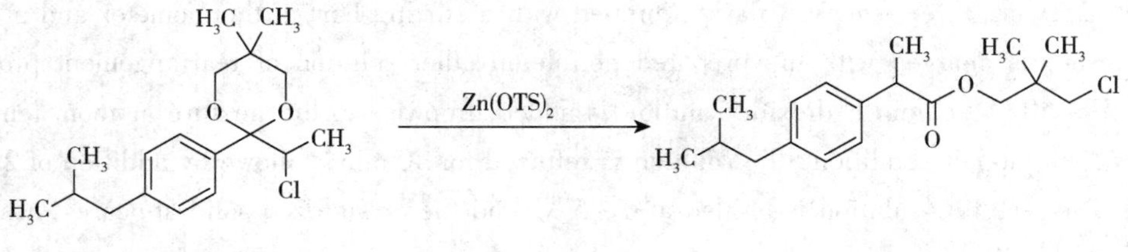

H_3C CH_3 O O CH_3 CH_3 Cl H_3C —Zn(OTS)$_2$→ CH_3 H_3C CH_3 O Cl CH_3 H_3C O

【Reagents】

2 – (1 – Chloroethyl) – 2 – (4 – isobutylphenyl) – 5, 5 – dimethyl – 1, 3 – dioxane: ketal – petroleum ether solution obtained in last step

Bis (*p* – toluenesulfonic acid) zinc: 0.6 g

Diatomite: 0.5 g

【Procedure】

A 250 ml, three – necked flask equipped with a stirring bar, a thermometer and an atmospheric distillation unit, is charge with the prepared ketal – petroleum ether solution and 0.6 g of bis (*p* – toluenesulfonic acid) zinc. The petroleum ether is recovered by distillation under stirring. Then, the distilling apparatus is replaced with a reflux unit. The mixture is allowed stirring at 140 ~ 150 ℃ for 2. 5 h and chilled to room temperature, at which point 0. 5 g of diatomite and 10 mL of petroleum ether are added. The resulting mixture is stirred for 0. 5 hand filtered. The filter cake is washed by petroleum ether (10 ml × 2) . The filtrate and washings are combined in a clean flask and can be directly used in the next step of reaction without further purification.

【Subjects for Thinking】

1. What is the mechanism of this reaction?
2. What is the reason for adding bis (*p* – toluenesulfonic acid) zinc into the reaction?

four The preparation of Ibuprofen

【Aim】

To comprehend the mechanism, reaction conditions and characteristics of etherification.

【Reaction equation】

CH_3 CH_3 H_3C CH_3 O Cl O H_3C $\xrightarrow[\text{2)}H^+]{\text{1)NaOH}}$ CH_3 CH_3 H_3C COOH + H_3C CH_3 HO Cl

【Reagents】

Sodium hydroxide solution (50 %): 3. 7 ml

Hydrochloric acid: 3 ml

【Procedure】

A 250 ml, three – necked flask equipped with a stirring bar, a thermometer and a reflux condenser, is charged with the prepared petroleum ether solution of rearrangement product. 3. 7 ml of 50 % sodium hydroxide solution is added dropwise to the mixture at room temperature. After complete addition, the solution is refluxed for 40 min, followe by addition of 2 ml of water. The resulting solution is chilled at 0 ~ 5 ℃ and the crystallized solid appeares, which is collected by filtration and washed with petroleum ether (10 ml × 2) . Therefore, the crude ibuprofen sodium (about 11. 0 g) is obtained.

The crude ibuprofen sodium salt is dissolved in 8 times volume of water, heated to reflux to complete dissolution, and charged with 0. 5 g of activated carbon for 30 minutes. The activated carbon is filtered off by a hot filtration and washed with 20 mL of water. The filtrate and washing are combined and charged with 3 mL of hydrochloric acid to adjuste the pH = 2. The crude product precipates at 0 ~ 5 ℃ and collected by filtration (about 8. 8 g) . If required, the fine product may be recrystallized from 70 % ethanol.

【Subjects for Thinking】

What is the operation principle of this reaction?

第三部分　设计性实验

实验二十三　吲哚美辛的合成及结构确证

【中文名称】 吲哚美辛

【英文名称】 Indometacin

【化学名】 2－｛1－［（4－氯苯基）甲酰基］－5－甲氧基－2－甲基－1*H*－吲哚－3－基｝乙酸 2－｛1－［（4－chlorophenyl）carbonyl］－5－methoxy－2－methyl－1*H*－indol－3－yl｝ acetic acid

【结构式】

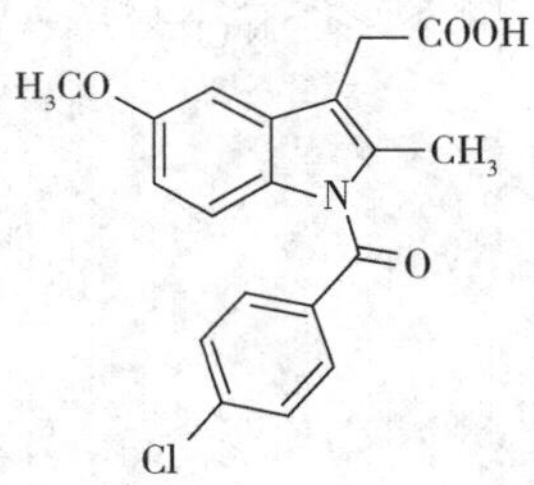

【分子式】 $C_{19}H_{16}ClNO_4$

【分子量】 357.79

【理化性质】 性状：白色或微黄色结晶性粉末，几乎无味，无臭，熔点155～162 ℃

【用途】 非甾体抗炎药

一、文献检索及合成路线设计

【实验目的】

1. 学习常用文献数据库的使用方法。
2. 学习化合物合成路线的设计方法。
3. 学习实验涉及的化学试剂的性质和使用方法。
4. 学习实验涉及的化学反应的反应机理。

【实验内容】

1. 根据吲哚美辛结构，通过 SciFinder、Thomson Integrity、CNKI 全文数据库和维普中文期刊等中外文数据库检索目标化合物的性质、合成路线和合成方法。

2. 根据相关文献，总结吲哚美辛的合成路线，并在实验教师的指导下，结合实验室现有条件，确定一条合理、可行的合成路线，制定实验方案，并确定每步操作的适宜的工艺条件，提出实验试剂及仪器需求。

3. 提前查阅和学习所用化学试剂的理化性质和使用注意事项。
4. 学习实验中涉及的化学反应机理。
5. 提交开题报告。

【思考题】

1. 在确定合成路线的过程中，主要考虑哪些因素？
2. 你所设计的路线与其余路线相比有什么特点？

二、对甲氧苯基重氮基磺酸钠的制备

【实验目的】

学习重氮盐制备反应及特点。

【反应式】

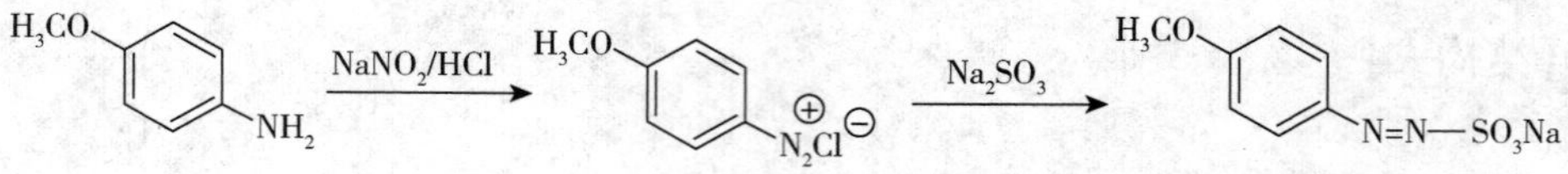

$$H_3CO\text{-}C_6H_4\text{-}NH_2 \xrightarrow{NaNO_2/HCl} H_3CO\text{-}C_6H_4\text{-}N_2^{\oplus}Cl^{\ominus} \xrightarrow{Na_2SO_3} H_3CO\text{-}C_6H_4\text{-}N{=}N\text{—}SO_3Na$$

【实验试剂】

对甲氧基苯胺：5 g
浓盐酸：10 ml
亚硝酸钠/水：2. 8 g /10 ml
亚硫酸钠/水：6. 5 g /25 ml

【操作步骤】

在配置有搅拌子和回流冷凝器的 250 ml 反应瓶中，依次加入 5 g 对甲氧基苯胺，22 ml蒸馏水和 10 ml 浓盐酸，冰水浴冷却至 10 ℃以下，搅拌下缓慢滴加亚硝酸钠水溶液（2. 8 g 亚硝酸钠 + 10 ml 水），滴加过程维持温度在 10 ℃以下，滴毕，在此温度下继续搅拌反应 30 min，用 30% 的氢氧化钠水溶液调反应液 pH = 8，搅拌下加入亚硫酸钠水溶液，加毕，室温下继续搅拌反应 1. 5 h，得对甲氧苯基重氮基磺酸钠溶液备用。

【注意事项】

1. 用淀粉 - 碘化钾试纸检测反应终点，如果未到反应终点可以适当延长反应时间。
2. 加入过程保持反应温度在 20 ~ 25 ℃。

【思考题】

1. 重氮化反应为什么要在低温下反应？
2. 重氮盐的反应液为什么要用氢氧化钠水溶液调 pH = 8？

三、对甲氧基苯肼磺酸钠的制备

【实验目的】

学习对甲氧基苯肼磺酸钠的制备及反应试剂的特点。

【反应式】

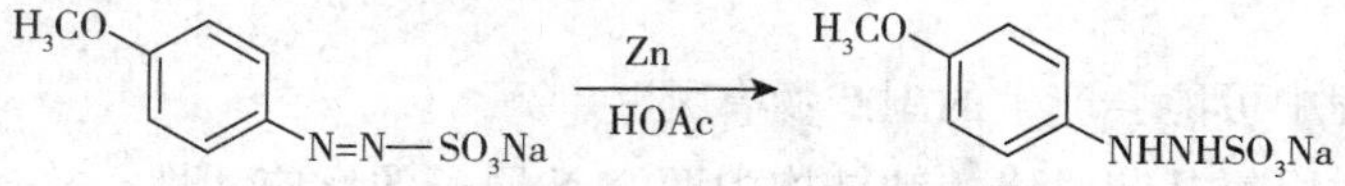

H_3CO–C_6H_4–$N{=}N{-}SO_3Na$ $\xrightarrow[HOAc]{Zn}$ H_3CO–C_6H_4–$NHNHSO_3Na$

【实验试剂】

甲氧苯基重氮基苯磺酸钠：前步反应制备量

冰醋酸：4 ml

锌粉：4 g

【操作步骤】

在配置有搅拌子和回流冷凝器的 250 ml 反应瓶中，加入制得的对甲氧苯基重氮磺酸钠和 4 ml 冰醋酸，升温至 40 ℃，加入 4 g 锌粉，搅拌 30 min，再升温至 45 ℃，继续搅拌反应 1. 5 h，趁热过滤，滤液冷却后析出大量白色固体，抽滤，冰水洗涤滤饼，干燥，得对甲氧基苯肼磺酸钠粗品。

【思考题】

本实验中的重氮基是否可以采用其他的还原方法?

四、N^1 – 对甲氧苯基 – 对氯苯甲酰肼的制备

【实验目的】

了解肼的 *N* – 酰化方法及特点。

【反应式】

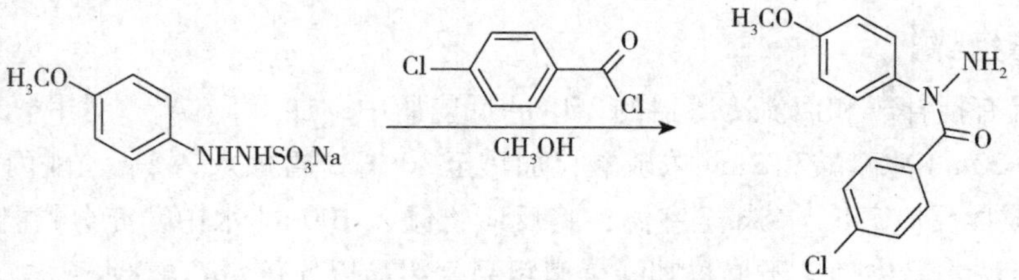

H_3CO–C_6H_4–$NHNHSO_3Na$ $\xrightarrow[CH_3OH]{Cl–C_6H_4–COCl}$ H_3CO–C_6H_4–$N(NH_2)$–CO–C_6H_4–Cl

【实验试剂】

对甲氧基苯肼磺酸钠：8 g

无水甲醇：25 ml

对氯苯甲酰氯：5. 8 g

氢氧化钠/水/乙醇：2. 4 g/10 ml/10 ml

【操作步骤】

在配置有搅拌子的 100 ml 反应瓶中，加入 8 g 对甲氧基苯肼磺酸钠和 24 ml 水，室温下搅拌 15 min，再加入无水甲醇 25 ml，于室温、搅拌下滴加 5. 8 g 对氯苯甲酰氯，加毕，继续搅拌反应 4 h，加入由氢氧化钠/水/乙醇（2. 4 g/10 ml/10 ml）组成的混合液，

室温下搅拌 20 min，抽滤，水洗滤饼，干燥，得 N^1－对甲氧苯基－对氯苯甲酰肼粗品。

【思考题】

1. 实验过程中可能会产生的副产物有哪些？

2. 试比较对甲氧基苯肼磺酸钠结构中两个 N 原子的反应活性。

五、吲哚美辛的合成

【实验目的】

学习利用 Fischer 法合成吲哚环的机理及反应特点。

【反应式】

H_3CO　NH_2　N　O　Cl　H_3C　O　OH　O　$H_2SO_4/HOAc$　COOH　H_3CO　CH_3　N　O　Cl

【实验试剂】

N^1－对甲氧苯基－对氯苯甲酰肼：5 g

乙酰丙酸：2. 2 g

冰醋酸：20 ml

浓硫酸：2 ml

【操作步骤】

在配置有搅拌子和回流冷凝器的250 ml 反应瓶中，加入 5 g N^1－对甲氧苯基－对氯苯甲酰肼、20 ml 冰醋酸和 2 ml 浓硫酸，加热至 30 ℃，再加入 2. 2 g 乙酰丙酸，升温至 40～50 ℃搅拌反应 3 h，冷却至室温，将反应物倾入 100 ml 水中，充分搅拌后，抽滤，水洗滤饼至洗水呈中性，干燥得吲哚美辛粗品。粗品以 5 倍量乙醇/水（1：1）重结晶 2 次，得吲哚美辛精品。

【思考题】

1. Fischer 法合成吲哚环的机理是什么？

2. 加入冰醋酸的目的是什么？是否可以用其他试剂替代？

六、产物的结构确证及纯度检测

【实验目的】

1. 学习用红外光谱法确证产物结构。

2. 学习用熔点法及液相色谱法检测产品纯度的方法。

【实验仪器】

红外分光光度计，质谱仪，压片机，显微熔点测定仪，液相色谱仪，精密天平

【实验试剂】

溴化钾（A. R.），甲醇（A. R.），冰醋酸（A. R.），乙腈（色谱纯）

【操作步骤】

1. 溴化钾压片：取溴化钾 200 mg 左右，置于干净的研钵中，在红外灯下充分研磨，用压片机压成厚度 1 ~ 2 mm 的透明晶片，在红外光谱仪上扫描，作为空白对照。

2. 制备产物薄片：取吲哚美辛 1 ~ 2 mg，溴化钾 200 mg，置于干净的研钵中，在红外灯下充分研磨，用压片机压成厚度 1 ~ 2 mm 的透明晶片，进行红外扫描，扣除溴化钾片的空白，得产物的红外光谱图。

3. 对图谱的主要吸收峰进行归属，并与标准图谱比对。

4. 显微熔点测定仪测定产物的熔点。吲哚美辛的 m. p 158 ~ 162 ℃。

5. 取吲哚美辛样品，进行质谱测试，测得其分子离子峰，判断产物分子量。

6. 取吲哚美辛约 25 mg，精密称定，置 50 ml 量瓶中，加甲醇适量，超声使溶解，放冷，用甲醇稀释至刻度，摇匀，精密量取 2 ml，置 10 ml 量瓶中，用 50 % 甲醇溶液稀释至刻度，摇匀，精密量取 20 μl，注入液相色谱仪，以乙腈 – 0. 1 mol/L 冰醋酸溶液(50 : 50) 为流动相；检测波长为 228 nm，记录色谱图。

7. 取吲哚美辛对照品约 25 mg，精密称定，置 50 ml 量瓶中，加甲醇适量，超声使溶解，放冷，用甲醇稀释至刻度，摇匀，精密量取 2 ml，置 10 ml 量瓶中，用 50 % 甲醇溶液稀释至刻度，摇匀，同法测定，按外标法以峰面积计算。

8. 完成综合以上步骤的结论，完成实验报告。

【注意事项】

1. 溴化钾粉末要充分干燥。

2. 根据所测产物的熔程可大致判断产物的纯度，一般熔程在 2 ℃范围内，可认为样品较纯；液相色谱法可定量判断纯度。

3. 若没有吲哚美辛对照品，可用面积归一化法计算吲哚美辛的纯度。

【思考题】

1. 吲哚美辛的红外图谱有何特点?

2. 吲哚美辛的质谱有哪些特征峰，有无杂质峰，其来源是什么?

Experiment 23 The synthesis of Indometacin and its structural verification

【Name】 Indometacin

【Chemical Name】 2 – {1 – [(4 – chlorophenyl) carbonyl] – 5 – methoxy – 2 – methyl – 1*H* – indol – 3 – yl} acetic acid

【Structure】

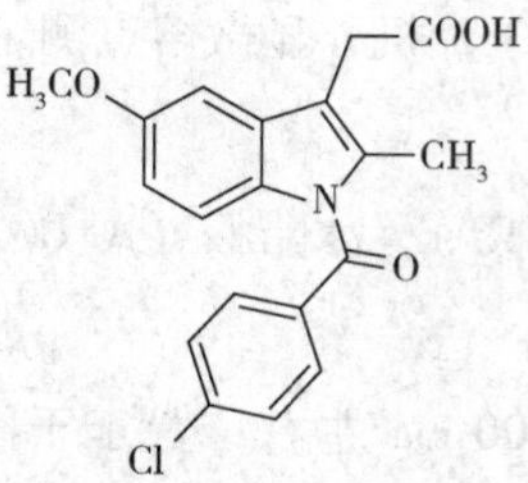

【Molecular Formula】$C_{19}H_{16}ClNO_4$

【Molecular Weight】357. 79

【Properties】White or light yellow crystalline powder, barely odorless, m. p 155 ~ 162 ℃

【Application】Non – steroidal anti – inflammatory drug

one　Literature retrieval and synthetic route design

【Aim】

1. To learn how to utilize literature databases.
2. To comprehend the design methods of synthetic routes of chemical compounds.
3. To learn the properties and the usage of reagents used in the experiment.
4. To comprehend the mechanism of reaction involved in the experiment.

【Content】

1. Search for the properties, synthetic routes and synthesis methods of the target compound based on the chemical structure of Indometacin using databases such as SciFinder, Thomson Integrity, CNKI or VIP.

2. Summarize the synthetic routes of Indometacin base on related literatures, design a rational and feasible synthetic route based on the current condition of the laboratory under the guidance of the tutors. Determine the experimental plan including technological conditions of each step of reactions and submit a manifest of the reagents and equipment required in the experiment.

3. Perform a preliminary search about the properties and items for caution of the chemical reagents that will be used in the experiment.

4. Learn the reaction mechanism of the reactions involved in the experiment.

5. Submit the research proposal.

【Subjects for thinking】

1. What are the factors that should be taken into consideration for the design of synthetic route?

2. What is the special feature of the route you designed comparing to other routes?

two The preparation of sodium 2 – (4 – methoxyphenyl) diazenesulfonate

【Aim】

To learn the preparation method and characteristics of the diazonium salts.

【Reaction equation】

$$H_3CO-C_6H_4-NH_2 \xrightarrow{NaNO_2/HCl} H_3CO-C_6H_4-\overset{\oplus}{N_2}\overset{\ominus}{Cl} \xrightarrow{Na_2SO_3} H_3CO-C_6H_4-N{=}N-SO_3Na$$

【Reagents】

p – Methoxyanline: 5 g

Concentrated hydrochloric acid: 10 ml

Sodium nitrite /Water: 2. 8 g /10 ml

Sodium sulfite /Water: 6. 5 g /25 ml

【Procedure】

A 250 ml, three – necked flask fitted with astrring bar and a reflux condenser is charged with 5. 0 g of *p* – methoxyanline, 22 ml of distilled water and 10 ml of concentrated hydrochloric acid. After the reaction system is cooled to 10 ℃ under ice water bath, a solution of 2. 8 g sodium nitrite in 10 ml of water is added dropwise slowly at such a rate that the internal temperature deos not exceed 10 ℃. The mixture is allowed stirring for additional 30 min, and subjected to 30 % sodium hydroxide to adjust the pH to 8. After the sodium sulfite solution (6. 5 g in 25 ml water) is added, the reaction is stirred under room temperature for 1. 5 h to give the sodium 2 – (4 – methoxyphenyl) diazenesulfonate solution for next step use.

【Notice】

1. A starch – potassium iodide testing strip is used to monitor the end of the reaction. Extend the reaction time if the test result is negative.

2. Maintain the internal temperature between 20 ℃ and 25 ℃ during the addition of sodium sulfite solution.

【Subjects for thinking】

1. Why diazotization reactions should be proceed under lower temperature?

2. Why the pH value of the diazonium salt solution should be adjusted to 8?

three The preparation of sodium 2 – (4 – methoxyphenyl) hydrazinesulfonate

【Aim】

To learn the preparation of sodium 2 – (4 – methoxyphenyl) hydrazinesulfonate and the properties of the reagents.

【Reaction equation】

$$H_3CO-C_6H_4-N{=}N-SO_3Na \xrightarrow[HOAc]{Zn} H_3CO-C_6H_4-NHNHSO_3Na$$

【Reagents】

Sodium 2 – (4 – methoxyphenyl) hydrazinesulfinate: obtained from last experiment

Glacial acetic acid: 4 ml

Zinc powder: 4 g

【Procedure】

A 250 ml, three – necked flask fitted with a strring bar and a reflux condenser is charged withthe solution of sodium 2 – (4 – methoxyphenyl) hydrazinesulfonate and 4 mL of glacial acetic acid. The mixture is stirred at 40 ℃, at which point 4. 0 g of zinc powder is added and the resulting mixture is stirred at 40 ℃ for 30 min and at 45 ℃ for 1. 5 h. After a hot filtration, the filtrate is cooled to room temperature, and a white solid precipitates, which is collected by filtration and washed with water.

【Subjects for thinking】

Is there any other reduction method for the diazo group?

four The preparation of 4 – chloro – *N* – (4 – methoxyphenyl) benzohydrazide

【Aim】

To learn the *N* – acylation method of hydrazine and its properties.

【Reaction equation】

$$H_3CO-C_6H_4-NHNHSO_3Na \xrightarrow[CH_3OH]{Cl-C_6H_4-COCl} H_3CO-C_6H_4-N(NH_2)-C(=O)-C_6H_4-Cl$$

【Reagents】

p – Methoxy phenylhydrazine sodium sulfonate: 8 g

Anhydrous methanol: 25 ml

p – Chlorobenzoyl Chloride: 5. 8 g

$NaOH/H_2O/EtOH$: 2. 4 g/10 ml/10 ml

【Procedure】

A 100 ml, round – bottomed flask fitted with a strring bar is charged with a solution of sodium 4 – methoxy phenylhydrazine sodium sulfonate (8.0 g) in water (24 ml). The mixture is stirred at room temperature for 15 min, and anhydrous methanol (25 ml) is added. Then, 4 – chlorobenzoyl chloride (5.8 g) is added dropwise, and the resulting mixture is allowed stirring at room temperature for 4 h. After the addition of the mixture of $NaOH/H_2O/C_2H_5OH$ (2.4 g/10 ml/10 ml), the stirring is continued for 20 min. The crude 4 – chloro – *N* – (4 – methoxyphenyl) – benzohydrazide is collected by filtration and washed with water.

【Subjects of thinking】

1. What are the by – products that could be generated in the procedure?

2. Try to compare the reactivity of the two nitrogen atoms in *p* – methoxy phenylhydrazine sodium sulfonate.

five The Synthesis of Indometacin

【Aim】

To learn the mechanism of forminga indole ring using Fischer reaction and its reaction characteristics.

【Reaction equation】

【Reagents】

N^1 – *p* – methoxylphenyl – 4 – chlorobenzoyl hydrazine: 5.0 g
Levulinic acid: 2.2 g
Glacial acetic acid: 20 ml
Concentrated sulfuric acid: 2 ml

【Procedure】

A 250 ml, three – necked flask fitted with astrring bar and a reflux condenser is charged with N^1 – *p* – methoxylphenyl – 4 – chlorobenzoyl hydrazine (5.0 g), glacial acetic acid (20 ml) and *con.* sulfuric acid (2 ml). The resulting mixture is stirred at 30 °C, then levulinic acid (2.2 g) is added. The resulting solution is allowed stirring at 40 ~ 50 ℃ for 3 h, and then poured into 100 ml of water to result in precipates under a fully stirring. The crude product is collected by filtration and washed with water until the washings are free of acid. The crude product is recrystallized from C_2H_5OH/H_2O (5 times to the mass of the crude) twice, giving the pure Indometacin.

【Subjects of thinking】

1. What is the mechanism of Fisher Indole reaction?

2. What is the purpose for adding glacial acetic acid? Are there any other reagents can be used instead of glacial acetic acid?

six The verification of the structure of target compound and purity determination

【Aim】

1. To learn how to verify the structure of the target compound utilizing infrared spectroscopy.

2. To learn how to determine the purity of the target compound using melting point method and liquid chromatography.

【Apparatus】

Infrared Spectrophotometer, Mass Spectrometer, Micro Melting Point Apparatus, Liquid Chromatograph, Analytical balance

【Reagents】

Potassium bromide (A. R.), Methanol (A. R.), Glacial acetic acid (A. R.), Acetonitrile (optical pure)

【Procedure】

1. Potassium bromide tableting: To a clean mortar, 200 mg potassium bromide is added and fully grinded under infrared light, then tableted into a 1 – 2 mm transparent crystal tablet using a tablet press. The tablet is then scanned on the infrared spectrometer and used as a blank control.

2. The preparation of the product tablet: To a clean mortar, 200 mg potassium bromide, 1 ~ 2 mg of Indometacin is added and fully grinded under infrared light, then tableted into a 1 ~ 2 mm transparent crystal tablet using a tablet press. Then tablet is then scanned on the infrared spectrometer, the infrared spectrogram is obtained after the deduction of the signal created by the potassium bromide tablet.

3. Confirm main peaks in the spectrum, and then compare with a standard spectrum.

4. Measure the melting point of the product using a micro melting point apparatus. the mp of Indometacin is 158 ~ 162 ℃.

5. A sample of the product is recorded on a MS instrument. The molecular weight of the products can be confirmed based on the molecular ion peak of the sample.

6. A sample of 25 mg of Indometacin is weighed precisely, then added to a 50 – ml measuring flask, along with appropriate amount of methanol. The sample is then dissolved under ultrasound. The Indometacin sample is diluted to scale with methano 1. 2 ml of the diluted sample is transferred to a 10 ml measuring flask, diluted to scale with 50 % methanol solution, then a 20 μl sample is inject into LC for testing. Acetonitrile – 0. 1 mol/L glacial acetic acid (50 : 50) are chosen as the mobile phase; the test wavelength is set to 228nm, record the spectrum.

7. A sample of 25 mg of Indometacin is weighed precisely, then added to a 50 ml measuring flask, along with appropriate amount of methanol, the sample is then dissolved under ultrasound.

The Indometacin sample is diluted to scale with methanol 2 ml of the diluted sample is transferred to a 10 ml measuring flask, and diluted to scale with 50 % methanol solution. Use the same measuring method as stated above, then calculated the content of Indometacin using external standard method based on the peak area.

8. Complete the experiment report after drawing a conclusion.

【Notice】

1. Potassium bromide powder should be fully desiccated.

2. The purity of the product can be judged based on its melting range. The sample can be considered pure if its melting range is within 2 degrees; liquid chromatography can be used to judge its purity quantitatively.

3. The sample's purity can be calculated using area normalization method if no Indometacin reference is available.

【Subjects of thinking】

1. What is the feature of the infrared spectrum of Indometacin?

2. What are the characteristic peaks in the mass spectrum of Indometacin? Are there any artificial peaks? If so, what are their sources?

附　录

附录一　压力单位换算表

	Pa	bar	kgf/cm²	atm	at	Torr	mm H_2O	mmHg	Psi
1 Pa（帕）	1	0.00001	0.00001	0.00001	0.00001	0.0075	0.10197	0.0075	0.00014
1 bar（巴）	100000	1	1.01972	0.9869	1.01972	750.062	10.1972	750.062	14.504
1 kgf/cm²	98066.5	0.98067	1	0.9678	1	735.6	10.000	735.6	14.22
1 atm（标准大气压）	101325	1.01325	1.033	1		760	10.332	760	14.7
1 at（工程大气压）	98067	0.98067	1	0.9678	1	735.6	10.000	735.6	14.22
1 Torr（托）	133.3	0.00133	0.00136	0.00132	0.00136	1	13.6	1	0.01934
1 mm H_2O（毫米水柱）	9.8067	0.000098	0.0001	0.0000968	0.0001	0.07356	1	0.07356	0.00142
1 mmHg（毫米汞柱）	133.322	0.00133	0.00136	0.00132	0.00136	1	13.5951	1	0.01934
1 Psi（磅/寸²）	6894.76	0.06895	0.07031	0.06805	0.07031	51.7149	703.07	51.7149	1

注：毫米水柱是指4℃状态的水柱高度，毫米汞柱是指0℃状态的水银柱高度。

附录二　干燥剂使用指南

干燥剂	适合干燥的物质	不适合干燥的物质	吸水量（g/g）	活化温度
氧化铝	烃，空气，氨气，氩气，氦气，氮气，氧气，氢气，二氧化碳，二氧化硫		0.2	175 ℃
氧化钡	有机碱，醇，醛，胺	酸性物质，二氧化碳	0.1	
氧化镁	烃，醛，醇，碱性气体，胺	酸性物质	0.5	800 ℃
氧化钙	醇，胺，氨气	酸性物质，酯	0.3	1000 ℃
硫酸钙	大多数有机物		0.066	235 ℃
硫酸铜	酯，醇，（特别适合苯和甲苯的干燥）		0.6	200 ℃
硫酸钠	氯代烷烃，氯代芳烃，醛，酮，酸		1.2	150 ℃
硫酸镁	酸，酮，醛，酯，腈	对酸敏感物质	0.2 0.8	200 ℃
氯化钙	氯代烷烃，氯代芳烃，酯，饱和芳香烃，芳香烃，醚	醇，胺，苯酚，醛，酰胺，氨基酸，某些酯和酮	0.2（$1H_2O$） 0.3（$2H_2O$）	250 ℃

续表

干燥剂	适合干燥的物质	不适合干燥的物质	吸水量（g/g）	活化温度
氯化锌	烃	氨，胺，醇	0.2	110 ℃
氢氧化钾	胺，有机碱	酸，苯酚，酯，酰胺，酸性气体，醛		
氢氧化钠	胺	酸，苯酚，酯，酰胺		
碳酸钾	醇，腈，酮，酯，胺	酸，苯酚	0.2	300 ℃
金属钠	饱和脂肪烃和芳香烃 烃，醚	酸，醇，醛，酮，胺，酯，氯代有机物，含水过高的物质		
五氧化二磷	烷烃，芳香烃，醚，氯代烷烃，氯代芳烃，腈，酸酐，腈，酯	醇，酸，胺，酮，氟化氢和氯化氢	0.5	
浓硫酸	惰性气体，氯化氢，氯气，一氧化碳，二氧化硫	基本不能与其他物质接触		
硅胶（6-16 目）	绝大部分有机物	氟化氢	0.2	200~350 ℃
分子筛（3A）	分子直径>3 A	分子直径<3 A	0.18	117~260 ℃
分子筛（4A）	分子直径>4 A	分子直径<4 A，乙醇，硫化氢，二氧化碳，二氧化硫，乙烯，乙炔，强酸	0.18	250 ℃
分子筛（5A）	分子直径>5 A，如，支链化合物和有 4 个碳原子以上的环	分子直径<5 A，如，丁醇，正丁烷到正 22 烷	0.18	250 ℃

附录三　用于有机液体较强的去水剂

干燥剂	与水形成的化合物	注 释
Na	NaOH，H_2	用于烃和醚的去水很出色；不得用于醇和卤代烃
CaH_2	Ca（OH）$_2$，H_2	最佳去水剂之一；比 $LiAlH_4$ 缓慢但效率高相对较安全。用于烃，醚，胺，酯，C4 和更高级的醇（勿用于 C1，C2，C3 醇），不得用于醛和活泼羧基化合物
$LiALH_4$	LiOH，Al（OH）$_3$，H_2	只使用于惰性溶剂［烃基，芳基卤（不能用于烷基卤），醚］；能与任何酸性氢和大多数功能团（卤基，硝基等）反应。使用时要小心；多余者可慢慢加入乙酸乙酯加以破坏
BaO 或 CaO	Ba（OH）$_2$或 Ca（OH）$_2$	慢而有效；主要适用于醇类和醚类，但不易用于对强碱敏感的化合物
P_2O_5	HPO_3，H_3PO_4，$H_4P_2O_7$	非常快而且效率高，高度耐酸，建议先预干燥．仅用于惰性化合物（尤其适用于烃，醚，卤代烃，酸，酐）

附录四 一些溶剂与水形成的二元共沸物

溶剂	沸点/℃	共沸点/℃	含水量/%	溶剂	沸点/℃	共沸点/℃	含水量/%
三氯甲烷	61.2	56.1	2.5	甲苯	110.5	85.0	20
四氯化碳	77.0	66.0	4.0	正丙醇	97.2	87.7	28.8
苯	80.4	69.2	8.8	异丁醇	108.4	89.9	88.2
丙烯腈	78.0	70.0	13.0	二甲苯	137~140.5	92.0	37.5
二氯乙烷	83.7	72.0	19.5	正丁醇	117.7	92.2	37.5
乙腈	82.0	76.0	16.0	吡啶	115.5	94.0	42
乙醇	78.3	78.1	4.4	异戊醇	131.0	95.1	49.6
乙酸乙酯	77.1	70.4	8.0	正戊醇	138.3	95.4	44.7
异丙醇	82.4	80.4	12.1	氯乙醇	129.0	97.8	59.0
乙醚	35	34	1.0	二硫化碳	46	44	2.0
甲酸	101	107	26				

附录五 常见有机溶剂间的共沸混合物

共沸混合物	组分的沸点/℃	共沸物的组成（质量）/%	共沸物的沸点/℃
乙醇－乙酸乙酯	78.3，78.0	30：70	72.0
乙醇－苯	78.3，80.6	32：68	68.2
乙醇－三氯甲烷	78.3，61.2	7：93	59.4
乙醇－四氯化碳	78.3，77.0	16：84	64.9
乙酸乙酯－四氯化碳	78.0，77.0	43：57	75.0
甲醇－四氯化碳	64.7，77.0	21：79	55.7
甲醇－苯	64.7，80.4	39：61	48.3
三氯甲烷－丙酮	61.2，56.4	80：20	64.7
甲苯－乙酸	101.5，118.5	72：28	105.4
乙醇－苯－水	78.3，80.6，100	19：74：7	64.9

附录六 实验室常用酸、碱的浓度

试剂名称	密度（20℃）g/ml	浓度 mol/L	质量分数
浓硫酸	1.84	18.0	0.960
浓盐酸	1.19	12.1	0.372
浓硝酸	1.42	15.9	0.704
磷　酸	1.70	14.8	0.855

续表

试剂名称	密度（20℃）g/ml	浓度 mol/L	质量分数
冰醋酸	1.05	17.45	0.998
浓氨水	0.90	14.53	0.566
浓氢氧化钠	1.54	19.4	0.505

注：表中数据摘自 Dean. JA Lange's Handbook of Chemistry. 13th edition. 1985

附录七　沈阳药科大学实验室管理相关制度

学生实验守则

1.《学生实验守则》是实验教学环节中对学生的一种规范和要求，凡进入实验室进行实验的学生都必须严格遵守。

2. 实验前，学生应认真阅读实验讲义和实验指导书，明确实验目的，熟知实验原理、步骤和实验所需仪器设备、药品、试剂和材料。

3. 学生上实验课不得迟到，不能参加实验的学生，要有所在系（部、院）签章的请假单。

4. 实验时应穿白大衣。要保持实验室安静，保持室内卫生和整洁，不得将与实验无关的任何物品带入实验室。

5. 实验时要严格听从教师指导，遵守实验室各项规章制度和仪器设备操作规程，未经允许不得动用实验室任何物品，不得随意走动；不得擅自离开实验室。

6. 实验时要注意安全，要特别注意防火、防腐蚀及自身防护，发现事故苗头要及时报告指导教师。

7. 实验中要爱护仪器设备，节约使用实验材料和药品、试剂对易损的实验用品，实验前后要仔细检查，发现丢失损坏要及时报告，不得自行处理。

8. 实验后要认真检查水、电、煤气，清扫实验室，所有实验用物品都要经实验室工作人员检查无误、无损后，方可离开实验室。

9. 要认真总结分析实验数据，完成实验报告。

沈阳药科大学教学实验室安全管理办法

第一章　总则

第一条　教学实验室安全管理是确保实验教学工作正常进行的前提保证，为加强教学实验室的安全管理，确保全校师生员工的人身和财产安全，特制定本管理办法。

第二条　教学实验室安全管理工作必须贯彻“安全第一、预防为主”的方针，坚持“谁主管、谁负责”和“谁使用、谁负责”的原则，建立健全安全管理长效机制，实现学校教学实验室安全有效的运行。

第三条　本办法适用于学校承担实验、实践教学任务的各实验教学中心、外语语音室、文献检索室、CAI 教室等。

第二章　教学实验室安全管理工作职责

第四条　各学院负责安全工作的领导应对教学实验室安全工作实施监督管理，负领导责任。

第五条　各教学实验室负责人是本教学实验室安全管理的第一责任人，其对本教学实验室安全管理负领导与监督责任，应定期组织有关人员进行安全教育与培训。

第六条　教学实验室应设置专（兼）职安全员，安全员应经过培训，具备一定的安全知识和技能。安全员负责本实验室的日常安全管理工作，负责督促实验人员遵守有关安全生产规章制度和安全操作规程。

第七条　教学实验室各类人员应履行工作场所和工作岗位规定的安全职责，对自己所在岗位的作业行为负直接责任。

第八条　教学实验室要对首次进行实验操作的人员进行安全教育和培训，在掌握各项实验室安全管理办法和基本知识，熟悉各项操作规程后，方可开始实验操作。

第九条　教学实验室要结合本实验室的具体情况，制定相应安全管理细则和仪器设备的安全操作规程，明确安全责任人，并悬挂公示。

第十条　教学实验室应积极宣传、普及一般急救知识和技能，如烧伤、创伤、中毒、触电等急救处理办法。

第十一条　教学实验室要每天对各实验室进行常规安全检查，还应定期（每月至少一次）对实验室进行全方位的安全检查，及时排除隐患。安全检查坚持自查与抽查相结合的原则，并建立实验室安全工作档案。

第三章　教学实验室安全管理工作

第十二条　教学实验室内的仪器设备、材料、工具等物品要摆放整齐，布局合理。各实验室应及时清理废旧物品，不堆放与实验室工作无关的物品，保证安全通道畅通。要严格做到“四防”即防火、防盗、防破坏、防灾害事故，“四关”即关门、窗、水、气，“一查”即检查仪器设备。

第十三条　教学实验室防火工作应以防为主，了解各类有关易燃、易爆物品知识及消防知识，杜绝火灾隐患。实验室防火工作的具体内容详见学校消防安全管理制度的有关规定。

第十四条　教学实验室要加强安全用电管理，不得擅自改装、拆修电器设施；不得乱接乱拉电线，实验室内不得有裸露的电线头；电源开关箱内不得堆放物品，以免触电或燃烧；使用高压动力电时，应穿戴绝缘胶鞋和手套，或用安全杆操作；有人触电时，应立即切断电源，或用绝缘物体将电线与人体分离后，再实施抢救。

第十五条　教学实验室在使用易燃、易爆、剧毒及细菌疫苗等危险品时，要严格按相关管理规定使用和保管，同时要有可靠的安全防范措施，并作好详细记录。

第十六条　教学实验室在涉及压力容器、电工、焊接、振动、噪声、高温、高压、辐射、强光闪烁、细菌疫苗及放射性物质的操作和实验时，要严格制定相关操作规程，落实相应的劳动保护措施。

第十七条　教学实验室在使用放射性物质时应避免放射性物质进入体内和污染身体；尽量减少人体接受外部辐射的剂量；尽量减少放射性物质扩散造成的危害；对放射性废物要储存在专用污物筒中并定期按规定处理。

第十八条　教学实验室在生物安全方面的管理工作应严格遵守国家《实验室生物安全通用要求》的相关规定。

第十九条　合理安排实验，节约使用实验动物，爱护实验动物。实验完成后，应将动物尸体包好后置于学校指定地点集中冷冻存放，由学校有关管理部门委托具有资质的单位进行集中统一处理。教学实验室确需饲育、观察实验动物的，应具备必要的专用房间、设备，具有符合要求的通风和清扫冲刷等条件，设置专人负责管理，防止由于管理不善而造成动物逃逸和环境污染。

第二十条　树立环境保护意识，废气、废物、废液的处理应进行分类收集，定点存放，由学校有关管理部门委托具有资质的单位进行集中统一处理。严禁将实验产生的可能污染环境的废液、废渣随便倒入水池或随意堆放填埋，违反规定的，对相关责任人进行批评教育，造成损失的应进行赔偿，造成严重后果的给予处分、罚款并通报批评。

第四章　教学仪器设备安全管理工作

第二十一条　教学实验室要根据仪器设备的性能要求，提供安装使用仪器设备的场所，做好水、电供应，并应根据仪器设备的不同情况落实防火、防潮、防热、防冻、防尘、防震、防磁、防腐蚀、防辐射等技术措施。

第二十二条　教学实验室必须制定仪器设备安全操作规程，使用仪器设备尤其是大型仪器设备的人员必须经过培训，考核合格后方可上岗。

第二十三条　教学实验室应定期对仪器设备进行维护、校验和标定。

第二十四条　仪器设备发生故障要及时组织维修，并做好维修记录。一般仪器设备的维修、拆卸需经教学实验室有关负责人同意，不具备维修专业知识的人员不得从事此项工作。大型仪器设备的维修主要依靠生产厂家及专门维修公司，一般不准自行拆卸。

第二十五条　要注意大型仪器设备的停电、停水保护，防止因电压波动或突然停电、停水造成仪器设备损坏。

第二十六条　教学实验室要根据仪器设备的性质配备相应的消防设备与器材，相关工作人员应学会正确使用，提高事故防范能力。

第二十七条　教学仪器设备安全工作要责任到人，仪器设备的管理人员是该仪器设备的安全负责人。仪器设备在使用过程中要有人管理，管理人员应经常进行安全检查，发现问题应及时向领导与主管部门报告并解决。

第二十八条　教学仪器设备不得外借给任何个人使用。

第二十九条　因责任事故造成教学仪器设备损坏或丢失的单位或个人，应按照仪器设备损坏丢失赔偿办法进行赔偿。

第五章　教学实验室安全事故处理

第三十条　教学实验室发生生产安全事故，当事人或事故现场有关人员应及时采取自救、互救措施，以减少人员伤亡和财产损失，并保护好事故现场，同时向学院领导和有关职能部门报告。

第三十一条　有关人员接到事故报告后应迅速组织抢救，防止事故扩大，并按事故报告规定如实上报事故情况，不得隐瞒、谎报或拖延不报，不得破坏事故现场和毁灭有关证据。

第三十二条　学校和各单位进行事故调查处理时，应按照实事求是的原则，查清事故原因，查明事故性质和责任。以书面形式报告事故情况，内容包括事故发生的经过和性质、事故发生的原因分析和责任、事故责任者的处理、事故教训以及群众接受教育的情况，采取的防范措施等。

第三十三条　根据事故大小、情节轻重，对事故肇事者和责任者按有关规定，给以相应的行政和经济处罚。

第六章　附则

第三十四条　本办法由教务处负责解释。

第三十五条　本规定自发布之日起施行，原《沈阳药科大学实验室安全管理办法》（沈药大校字〔2009〕15 号）同时废止。

参考文献

[1] Smith M B, Jerry M. March's Advanced Organic Chemistry [M]. Wiley - Interscience, 2013.

[2] 罗代暄，吴培成，杨国栋，等. 化学试剂与精细化学品合成基础（有机分册）[M]. 北京：高等教育出版社，1991.

[3] Zweifel G S, Nantz M H, Somfai P. Modern Organic Synthesis: An Introduction [M]. The Scripps Research Institute, 2017.

[4] 黄宪，王彦广，陈振初. 新编有机合成化学（第二版）[M]. 北京：化学工业出版社，2003.

[5] 叶非，黄长干，徐翠莲. 有机合成化学 [M]. 北京：化学工业出版社，2010.

[6] 马俊林，刘栓柱. 苯佐卡因的合成研究 [J]. 武汉：湖北工业职业技术学院学报，2001，14（2）：74 - 77.

[7] 于凤丽，赵玉亮，金子林. 布洛芬合成绿色化进展 [J]. 有机化学，2003，23（11）：1198 - 1204.

[8] 麻生明，朱灿. 多取代吲哚、合成方法及其应用于吲哚美辛的合成 [P]. 2013，CN103012241A.

[9] 李公春，田源，李存希，等. 硝苯地平的合成 [J]. 浙江化工，2015，46（3）：26 - 29.

[10] 王燕，沈大冬，朱锦桃. （S） - 和（R） - 普萘洛尔的不对称合成 [J]. 有机化学，2007，27（5）：678 - 681.